I0057644

MORNING STAR

FARAH ALQATTAN

Contents

WORDS FROM THE AUTHOR

Guardians and Caretakers,

Protecting the sacred essence of who and what we are has required patience, endurance, faith, and courage. No act in goodness, no matter how small, is ever wasted. Continue to be generous, considerate, and kind. One thing remains permanent, no matter the change, is who we truly are. Hold-fasten to integrity, love, and truth. No matter how hard the fight or the challenge, know, we have won all the battles and wars; should we lead with love.

You are cherished immensely, in heart and soul.

Never give up... never ever give up, on each other, no matter what darkness be.

You know why?

-Smiles-

Because there's plenty of life that still breathes.

Print Version: 2, 2025
Print ISBN: 979-8-9881785-7-6
Library of Congress Control Number: 2025912060
For permission requests, contact: rwhpublishing@gmail.com
Author Instagram: farahalqattanofficial

Rwh Publishing LLC produces creative content by artists who aim to uplift humanity and change lives for the better.

VI ~

1

Publication Objective

To eradicate and continuously remove any measures or means that exploit personhood, especially for the purpose of monetary gain manifesting from ignorance, violence, hate or greed. This can include but is not limited to porngraphy, prostitution, brothel operations, hyper sexualization messaging, mallegislation, intrusive and obscene media, malproducts, malservices and other facilitators and operations. This is to align personhood to vitality, vigor, flourishment, nourishment, longevity and prosperity. And to improve awareness and understanding as to establish true justice for all people.

***Disclaimer**: Findings of this project are based on experiential data, open source intel, community feedback, and publicly shared information. This includes and is not limited to the internet, books, on-foot field analysis, qualitative research and higher guidance. Results and conclusions are shared to expand awareness as to include pertinent matter of facts to improve decision making processes and reduce injustice, violence, conflict, and/or hostility. Under no circumstance does the author or publisher condone misinformation, malinformation, disinformation, hate, violence, and/or corruption. Farah Alqattan and Rwh Publishing LLC do not assume responsibility for actions taken by individuals based on the information provided, but support the upliftment, improvement, and flourishment of life.*

2

History

Being of reason, to distinguish, the innate requirements needed to act accordingly towards another is not therein founded on the laws distinguished by man but founded in response to moral conscience as it is manifested. While variables are present which dilute, misguide, or inhibit judgment and reason within a landscape or surface area; conscience and universal feedback regarding decisions or choices are made evident as to address misalignment and manifestations of hurtful or harmful outcomes. To identify optimal function, the law ought to be scoped and be in support of the development and conduct of subjects through the space time continuum.

Since 3500 B.C. laws, regulations, legislation, and the courts have escalated sexualization and indecency by trading soulhood functions for self-interests and self-determination. Self-concepts, impacting behavior, are presented to subjects in landscapes reflecting world operation while aligning to specified actions rather than optimal conduct and vital functions. This is due to unseen and seen forces, influences, misinterpretations, rationalizations, and justifications contributing to norms, culture, identities and amplified divides. Overtime, individuals have relied on the common sense and moral reason of another, rather than consulting their own conscience to make decisions. Thus, by following other mortal directives, persons may be susceptible to ignoring presented and manifested occurrences and therefore abide in ignorance. This has been presented in the escalated legalization of prostitution across countries, contributing to ailments, illness, and disorders due to subjects acting in critical risk capacities unknowingly.

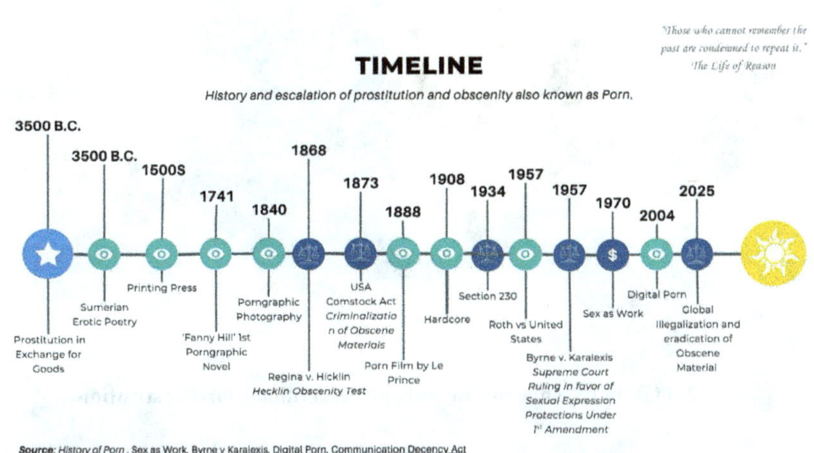

1.0 History of Personhood Exploitation, Represented by Prostitution

The global response to prostitution varies by country. It illuminates the inconsistencies and distortion, in reasoning and alignment, to justify personhood exploitation. All while allowing such acts for purposes contradictory to consciousness and optimal operation. As of 2025, 52% of the globe has legalized the sale of persons for the purpose of prostitution and 67% have legalized the purchase of persons for the purpose of sexual exploitation. Furthermore, most current global leadership requires guidance as to align in support of the dignity of soul experience, as to prevent negative manifestations of hurt outcomes, seen and unseen. Moreover, current exploitation mallegislation have contributed to the facilitation of maloperations. For example, operations presented as attractions for pleasure, amusement, entertainment, and/or tourism provide a controlled space for crime and exploitation. (Illic, 2023)

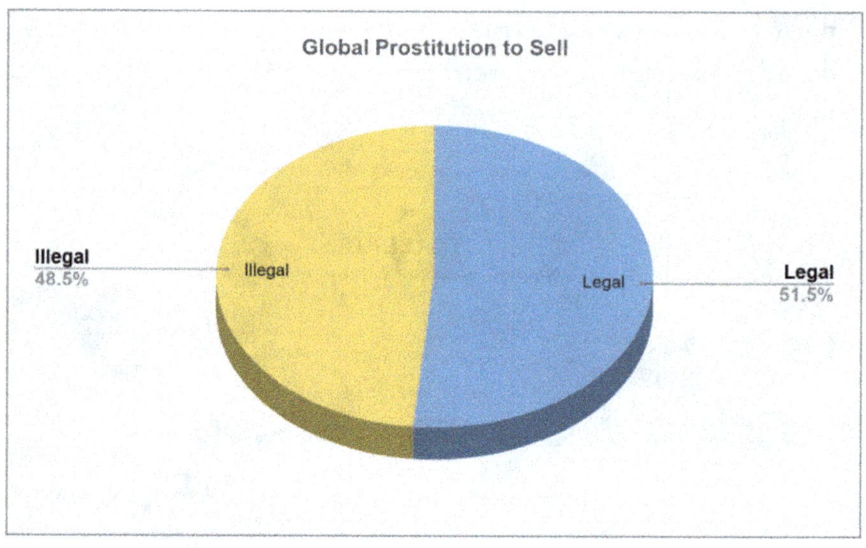

2.0 Global Stance on the Sale of Personhood for Prostitution

The ability to perceive and distinguish violations towards life, is not a wager to be taken lightly in mere arguments of debates as self imposing justifications and opinions for desire, as reflected in current democratic systems in law making. While this truth is challenging to digest, it is significant to integrate in an effort to improve current governance structures. The feedback for any conduct, surfacing, demands evaluation, reflection, and review. The overarching universal law is founded to support the abilities of each person to assess and respond to manifestations from personal conduct, as to align to responsibility. While perceptivity requires each to evaluate surfaced signals in experience, to improve conscious decision making and laws, each person is required to go beyond the presented baseline and function of environmental operations as to distinguish the standards for best conduct. Such alignment would be in congruence to universal awareness and learning as to improve judgment, understanding, and decision making.

While the United States prioritizes innovation, being able to distinguish the influence and impact of technology development may not be easily weighed as persons are living day to day and confined into scope of operations and function as it pertains to roles and responsibilities. Its speculative that varying data science technologies merely weigh scoped conditions and variables, based on area's of interest, like revenue retention or finances as to improve the State of the Union. For this reason, it is imperative to distinguish and size up consumption appetites to ensure the health and vitality of the populous general body is not devoured by machines or material things.

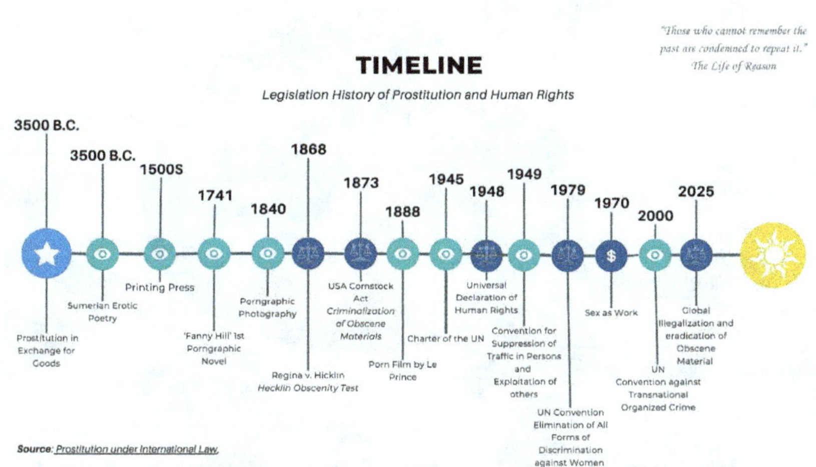

3.0 History of Prostitution and Human Rights within the U.S.

As it pertains to pornography, the United States leads all nations in the production of sexually explicit content. This may be due to the leniency in controls and standardization in the virtual and physical landscapes. The United States federal law has chosen to legalize prostitution, while all states have enacted illegalization laws, with the exception of Nevada, home of 'Sin' City. *(Monkey, 2025)* The stance each country has taken on exploitation has cultivated norms, manifesting subcultures, to later become projected and reinforced in media, television, social platforms, and printing. Within the United States, to combat sexually transmitted diseases and the sexualization of society, infrastructures like Planned Parenthood developed to advocate for clients' health while providing care taking resources. Moreover, other infrastructures have advocated for protected sex, as to deter and prevent harmful outcomes, by providing condoms and testing kits. As generations develop, they perceive access to spaces as norm(al) and integrate into lifestyle activity routines founded on the cultivated natural operations.

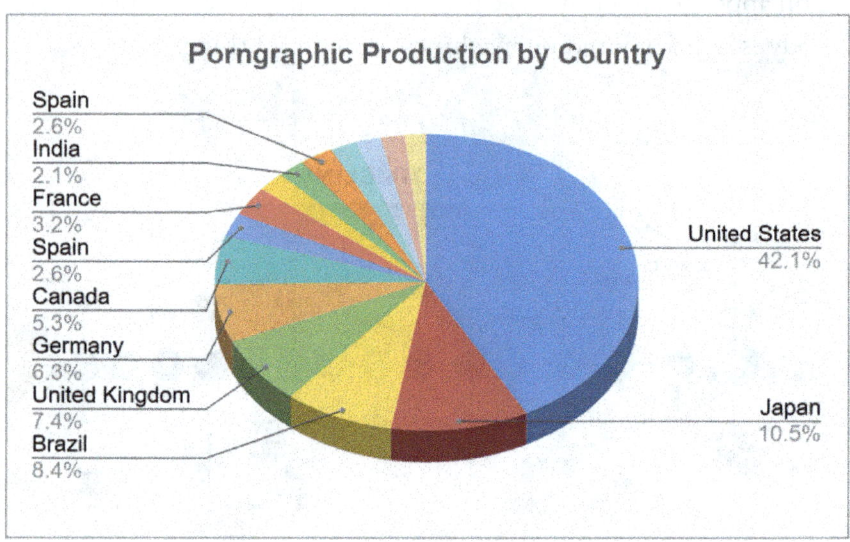

3.0 Countries Leading Production of Explicit Material Stimulating Arousal

There are various methods and areas, direct and indirect, persons occupy with the premeditated intention of obtaining, facilitating, or providing sexual acts. These locations and areas produce varying risk to the safety of the actors. For high risk locations, sex providers (victims) have disadvantages due to present elements inhibiting conscious decision making, like drugs and alcohol. Medium risk environments provide some level of control and mechanisms of peer support, as well as income predictability and protections due to healthcare resource access. Low risk operations are within protected venues, where partakers make limited contact and minimized incidence of penetrative acts. (*The Many Faces of Sex Work*, 2004) While the open space was founded to nourish and replenish personhood function; current operating businesses and high traffic spaces are scoped for pleasure and indulgence based consumption as to entertain and amuse. Furthermore, operations are perceived as 'fun' due to their immediate stimulating effects, including subculture in-group thinking and language towards risk taking like "Yolo" meaning "You only live once!" and "But, did you die?"

Direct	Indirect
Streets	Bondage Role Play
Brothels	Lap Dancing
Escorting	Massage Parlour
Private Grounds	Traveling Entertainment
Festivals	Street Vendors and Traders
Clubs or Pubs	Services of Gifts
Same-sex Venues	Arrangement
Hotels or Rentals	Swingers Club
Trucks, Trains, Cars, or Ships	Geisha
Activate CB Radio Signaling	Sex for Drugs
Call Services	Survival and Deprivation

4.0 Direct and Indirect Exploitative and/or Grooming Locations

3

Abstract

Current operations and functions produce seen and unseen, as well physical and virtual influences on subjects within the ecosystem. Consumption through sense processing triggers, enables, reinforces and propagates cyclic routines or habits leading to outcomes. This is further amplified by algorithms and advertisement targeting groups towards consumption based on mental states, social status, physiological health, identity demographics, age groups, and personhood needs. Moreover, physiologic inhibitors through consumption like alcohol and substances, legal or illegal, reduce optimal vital functionality and disrupt field coherence, reasoning, and sound judgment. This can lead to impulsivity, due to weakening neural system processing speeds triggering prompt action. For example, alcohol is often referred to as "liquid courage" indicating it's effects for reducing reasoning as to support impulsivity, characterized by impulse, or lack of impulse control . Moreover, innate and embedded personhood processes, like mirror neurons designed for empathy, can subconsciously affect a persons sense of relativity with peers, heightening conformity, agreeableness, or mimicry.

Researched surface area, is representative of San Francisco, Ashbury Haight District. Subjects access proximate available local establishments in neighborhoods and city landscapes as with other cities around the world. Digital identity or personas coupled with algorithmic operations due to data collection reinforce lifestyle due to personalized marketing and suggestive prompting. Moreover, other environmental stimuli can provide messaging as to support behavior and reinforce activity. For example, head banners and signs signal to subjects to act therein rather than prevent. Younger and vulnerable groups may be more susceptible to buy-in due excitement for experiential newness and 'fun' seeking for desired cognitive states and effects. For example, majority of participants questioned in the Haight District express they consume psychotropic products "to take the edge off" or "to help them relax". One consumer went as far as to say that his use gives him "peace of mind". Such rationalization by subjects map experiences to the immediate effects metabolized by physiological processes with achieved mental states that are euphoric, soothing, or numbing. When consumers mistaken effects, due to rationalization or justification, as having a 'sense' of peace, what they're actually communicating is experienced personhood stress or overwhelm. While many can achieve relief states through other means, due to culture, marketing, group think, rationalization, or available product offerings many resort for 'quick fixes'. Such fixes can represent needs for stimulation that derives feelings similar experienced with inclusion, belonging, alleviation or soothing, and even significance.

Some organizations producing substances and supplements with psychotropic controlled ingredients have been doing so under the disguises of 'medicinal' or 'plant based' with misleading messaging, while omitting short and long term consequences contributing to slue of harmful physiological outcomes like paranoia, confusion, and anxiety. Moreover, efforts to tackle crime by allocating continued funding to privacy violating innovation and detection, like drones or computer vision, will not address the root causes contributing to harms like violent protests or mental heath epidemics like loneliness experienced by young adults. As variables producing these harms, seen or unseen, are related to personhood consumption, negative thought patterns, absence of code of conduct or moral foundation, landscape architecture, availability of malproducts and malservices, mallegislation reinforcing maloperations, and misinterpretation due to ignorance or absent-mindedness. The research presented is practical and serves to reflect empathy as to showcase what it would be like to walk someone's shoes by tracing their footprint, as to provide constructive permanent resolution to existing societal problems. Moreover, this research is shared as to support the improvement of awareness and understanding as to prevent conflict that may be local or global violence, discrimination, prejudice, and/or bias.

4

Harms and Costs

Socially, personhood exploitation, like prostitution, can cause loss of basic human rights, loss of one's childhood, disruption in families, and severe mental health consequences. The psychological impact of body violations produces onsets of disorders like anxiety, post-traumatic stress disorder (PTSD), depression, loneliness and varying other cognitive disruptions that inhibit sound judgment and coherence. (Goldenberg et al., 2017) Other harms stemming from personhood violations produce higher levels of fear, suicide ideation, anxiousness, and memory loss. Moreover, victims exhibit Stockholm syndrome in which they grow attachment to their abusers. A multitude of physiological processes are disrupted due to sensory abuse, including but not limited to memory gaps, brain fog, bodily injuries, gastrointestinal complications, oral health decay, weight loss, and slew of other dysfunctions, disorders, and/or nervous system disregulation. Connection with peers and social support systems are traded with isolation due to shame, stigmatization, and other unseen derivatives from activities. This may erode the subjects sense of trust and safety; giving to illness and ailment due to negative thought patterns, rumination, and/or dis ease.

Spiritually, according to the Bible and Islam, prostitution is considered a consequence from persons not being in connection with

God. This is reflected in the lack of integrated value based guidance in decision making founded on virtues of uprightness. (*Topical Bible: Prostitution and Adultery*, n.d.) However, due to physiological dilution in sense making and operations of identity, persons engaging in harmful industries or areas are apt to mirror and model psychological thought patterns and resonance of energies in spaces. Moreover, persons operate with subdued judgment due to consumption of mind altering or numbing substances alternating mental state function. This is likely to inhibit or amplify reasoning processing speeds and contribute to impulsivity or worse risk taking behaviors as individuals operate under or with influence of persons, places, and things.

Criminal justice Routine Activity Theory distinguishes that offending takes place due to the availability of three elements. As described in routine activity theory, offenders partake in offending due to the lack of guardianship, perceived motivation, and availability of suitable target(s). (*Routine Activities Theory: Definition and Meaning*, 2021) While 67% of the globe has legalized the purchase of persons, giving the elements of motivation and allocating spaces for suitable target(s), signaling to offenders that 'it's acceptable and permissible' to act therein. Furthermore, with allocated business operations and peer influence, guardianship is replaced with advocacy and incitement while reflecting the architectural principle 'Form Follows Function.' Community established standards and baselines contribute to thoughts, likely framed from founded internet or media content, inclining decision making and standards of behavior . Overtime, this derives lifestyle routines contributing and reinforcing societal outcomes based on availability, status, state, and access. Moreover, some violators exploit scientific discoveries by developing mechanisms and means to bypass personhood defenses. This is reflected by grooming tactics in which offenders try to isolate victims as to develop dependency, influence, and thought infiltration for desired behavioral outcomes.

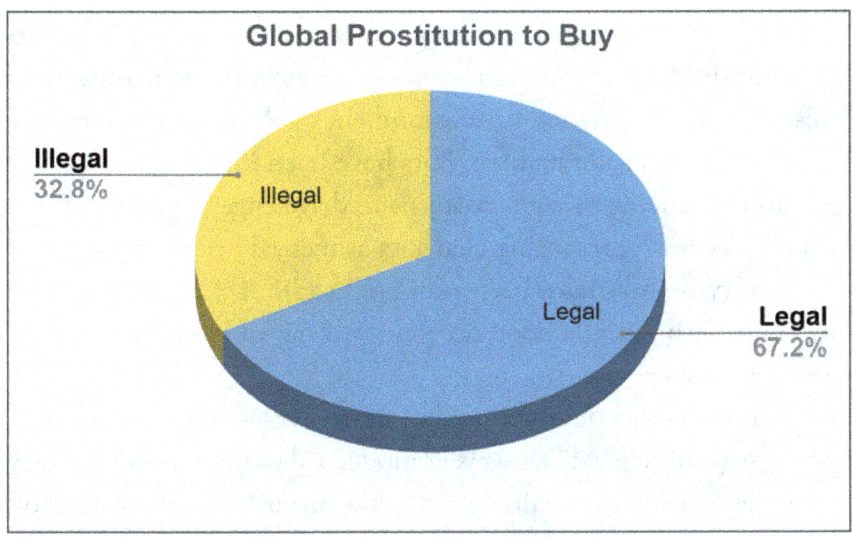

6.0 Global Stance on the Purchase of Personhood for Prostitution

Decriminalization of prostitution has provided opportunities to reinforce and enable personhood exploitation, especially through the virtual landscape. For example, website developers have built digital infrastructure that host and upload explicit content, while manifesting harmful industries like human trafficking and sexualization of cultures through developed software applications (apps). All while incentivizing risk taking behavior with rewards and promotions via operant conditioning and profiting by means of collecting data as to develop more targeted and personalized tools to manipulate and device persons towards consumption as to retain revenue and reinforce harmful cycles. Majority of stakeholders are not aware to company culture, projects, and priorities but operate in trust and faith while participating in the maintenance of the 'status quo'. Furthermore, some technologist founders deny and deflect accountability when questioned or subpoenaed in assembly hearings regarding algorithmic designs, that is currently impacting end-user experience. Yet, technology design mimics the neuroscience of a person, categorized as 'brain-inspired computing'. (Zhang, 2020) Technological developments negate the awareness of effects, short, and long term consequences on the ecosystem at large. Generations later, movements for 'Responsible AI' and 'AI Governance' are now being manifested, representing the continued disillusionment of the harms stemming from algorithmic decision making. Bots have been immersed in social platforms influencing thought patterns to drive engagement or fraud consumers by misrepresenting chatbots as mental health therapist. As of 2025, varying bills have been proposed to limit AI reach and infiltration impacting personhood decision making and inhibit source or god given guidance.

Every founder lays a foundation on which society operates and therefore influenced. However, modern day operation provides executives within the United States the means to deflect and bypass manifested outcomes through legal proceedings, litigation, arbitrary agreements, and the courts. For example, United States legislation Section 230 has been used as a prop to shield technology companies, like social platforms with embedded 'influencer' profile classification, from accountability and responsibility of their manifested impacts. Social platforms are often misclassified and miscategorized as "mirrors" as to deflect the reality in which they represent enticement, suggestion, prompting, and reinforcement through algorithmic operations. Numerous technology founders have been warned for years by public servants, in senate hearings, regarding the reckoning they would face for continuously dismissing, deflecting, and denying the evidence of the outcomes from platform designs that have impacted the populus' health and general body while deriving incivility, discord, indecency, and exploitation. (*Whistleblower Tells Senators That Meta Undermined US Security, Interests*, 2025) While average Americans pay a $1,000 for a mobile device, reflected by a weeks pay of an average worker, the actual cost of energy expended, converted to time, that's been wasted while 'Doom Scrolling' is unseen and may be not assessed. Furthermore, manifestation of psych disorders are attributed from divided attention with given device mobility that has interrupted attention processes and focus, mislead in reasoning, and impacted lived experiences of persons, teams, and family units.

Awareness of optimal function is recognized in education domains pertaining to psychology, sociology, neuroscience, philosophy, humanities and more. Yet, renowned scientists and professionals operate in structures exhibiting group-think and hive-mind with fear of sharing truth due retaliation, contracts, or bindings obligations interconnected with national security, family, legal status, and/or prestige. While citizens receive hundred dollar parking citations for not moving vehicles for street cleaning in California bay region, the business landscape in the virtual environment is yet to be cleaned up from its filth and be swiftly brought to justice for its impacts and harms. While business environmental impact assessment (EIA) permits are required for infrastructure development to ensure ecosystem balance and stability, current products and services in both the physical and digital landscapes are not optimized for flourishing; even if structures have been established through research and development programs, sting covert operations, or spiritual testing as to assess character. (*LCA: Evaluating Environmental Impact*, n.d.) To sum it up, Bright Hue famously once said, "In an effort to catch the dark, they propagated darkness onto the light." Thus, digital infrastructure founders that propagated and reinforced harms shall be penalized just as ordinary citizen have for their common mishaps.

Prostitution and personhood exploitation, stemming from justification, ignorance, greed, envy, and rationalization can cost more than just fiscal revenue, since expenses are continuously allocated to address short and long term consequences as it pertains to incident response, social work, policing, cyber security, insurance, and healthcare. Not to mention the re-lived generational trauma by survivors, in surface area's that refuse to change and align to higher purpose standardization. They've often said things are the way they are because 'sex sells' and currently each sex worker charges a range for their service, between 350-50 depending on the currency of the country they're in; a

1-hour service session in Colombia can cost 50 pesos which converts to 0.12 US dollars. (*The Price of Sex Around the World Mapped*, 2019) Thus, people of higher income in certain countries can utilize their wealth to exploit persons in less economically developed nations. Furthermore, persons in challenged or tourist communities do not question presented variables or the operation as it is presented to them, since it aligns with legislation and social norms. To meet and obtain income for survival, many end up resorting to operating through means to gain funding, including but not limited to seeking opportunities for quick cash like working in bars, brothels, stripping, prostituting, massage escorting, digital content creation of personhood compromising acts that reflect exposure of vulnerability and innocence. Such operations reinforce inequity considering persons develop a misrepresented sense of value regarding themselves and soul worth. History has always echoed the same message like that shared by Jessie Potter, "If you always do what you've always done, you always get what you've always gotten" or as expressed by Albert Einstein, "The definition of insanity is doing the same thing over and over and expecting different results."

Many citizens living in budget constraints may rationalize and justify participation in exploitation. It is important to recognize those categorized and classified as 'sinners' are actually persons, 'victims', basing their decision making process on their own understanding in context of their identity, environment, social status, mental state, and other seen and unseen variables. Children, like many who experience violations to their personhood can grow up to dismiss and bypass sound judgment due to learned behavior, associative peer pressure, economic need demands, inhibited states from mind altering consumption, and slue of other influences. Thus, it is imperative to take exploitation seriously, as a violation to soul operation, with a compassionate and dignified scope of analysis, as to improve awareness, education, and carve opportunities to support meaningful change that improves the livelihood of all people, across space time continuum.

The fiscal inequality between people has created a misconception within power dynamics founded on fiscal value rather than moral attributes, between buyers and sellers or offenders and victims, stemming from socioeconomic disparities as projected by national economies. This has influenced rationalization of persons in poverty as to function in lack mindset, causing victims to justify their actions and live with their oppressors and at times not being paid or paid very little. (Novotney, 2023) According to The Center for Child Protections, perpetrators often groom their victims over a duration of time, developing psychological, emotional, and/or financial dependency while exploiting victims' trust through manipulation. Child grooming tactics can include: (Munoz, 2019)

- Targeting kids for special attention, with offers of gifts or activities. (Conditioning)
- Slowly isolating a kid from family members and friends. (Dependency)
- Undermining relationships with parents and friends to show that "No one understands you like I do." (Grooming)
- Gradually pushing or crossing boundaries. This could range from long-lasting full-frontal hugs, sitting on laps or "accidental" touching of private parts. (Violating)
- Offenders engage in partially clothed tickle sessions, showering with kids or sleeping in the same bed. (Stimulation)
- Encourage kids to keep secrets from others. (Bonding and Tethering)

Varying marketing strategies hosted by exploiters aim to incentivize behavior toward risk taking through allocated funding or provide returns on investment from participation, engagement, or investment. This can be presented as 'lucrative opprotunities' but is a gross negligence, misrepresentation, and manipulation of true costs and harms, by means of carrot dangling 'capital gain' to entrap vulnerable persons. Thus, the true costs to combat harms far exceeds the minuscule rewards, due to the offset of expenses and consequences needed for education preventions, child protective services, detention facilities due to crime, information technology resources for monitoring, healthcare to address ailments, societal trust erosion, and much more.

Generationally, exposure to justifying decision-making can be mimicked by children through their lifespan. Thus, rationalization may be learned from established culture, not founded on conscious-based awareness processing. Psychiatrist Dr. William Glasser identified that decision making is purposeful with five aspects guiding behavior including love, freedom, power, fun, and survival. According to Dr. Glasser, teaching Choice Theory can instill empowerment and freedom in decision making thus supporting critical thought in weighing variables while considering alternatives.(Glasser, 2001) A person in any environment can consider their needs while weighing if the presented options to meet them in the landscapes are the only option, as to obtain funding, power, love, or exercise freedoms of expression or play. When individuals reflect and critically think in authenticity, they can distinguish their value represented by vitality and vigor, is far more significant than any gratification that can lead to disintegration. Persons may be best apt to be empowered to evaluate universal feedback from conduct, as surfaced signal manifestations, as to make better more informed and higher guided conscious-based decisions; irrespective of peer influence, opinions, operations, legislation, or technology; as founded in 'common wealth nations'.

While varying individuals are working day-to-day to combat harms fulfilling roles being caretakers and guardians, better legislation and cultural shifts can empower the cultivation of dignified and revered sentiment regarding experiences, especially as it pertains to intimacy. While current media and motion picture projects have presented varying personhood functions reflective of impulsivity, toxicity, and promiscuity; shifts to integrate value-based operation, aligned to sacredness, can improve culture to dignity and honor. Empowering individuals to take care of their health will likely reduce demand for healthcare. Moreover, findings from empirical evidence from feedback echoed in the breadth of survivor stories ought to be heard and valued, as precedents for better direction and policy making. A shift to acknowledgment, rather than deflection or denial, will elevate consciousness and awareness to process required stimuli for optimal interpenetration and improved decision making.

5

Benefits

The benefits for engaging with prostitution may include immediate financial reward, expanded social associations, physiological euphoria, expanded working network, access to resources, and other incentives and rewards as provided by the groomers and facilitators in a setting. According to the United States Department of Justice, "An analysis of prostitution in San Francisco, Nevada, Boston, and Dallas illustrates respectively the laissez faire, regulation, zoning, and control models. Under all of the models, prostitution will continue, but under varying incentives (money) and disincentives (criminal penalties)" (*Economics of Prostitution*, n.d.) This methodology of handling prostitution reflects the psychological model of operant conditioning, in which subjects are conditioned by the consequence of their behavior. (McLeod, 2025) Through this, individuals involved with prostitution may find the immediate reward from engaging in prostitution or sex trade as positive considering the immediate rewards gained. Overtime, as inverted and skewed as this is, developed habits by actors tether and establish dependencies that bondage victims to substances, perpetrators, and slue of other attachments that scale harms to peers, associations, and communities. Thus, utilizing incentivization models and methodologies do not deter behavior, but rather allocate, reinforce, and amplify manifestations due to deeper processes within the social, technological, psychological, financial and physiological domains of experiences. Furthermore, it amplifies and contributes to the drug trade as well as the human trafficking and organ harvesting industries.

Legalization arguments for harmful products or services merely presents an illusion of power and control, while contributing to injustice of vulnerable populations and the collective spirit. Looking at policies with a subset of variables, dismisses the evolving nature of technology, physiological functions, the psychological interpretation, and mental models influencing cognitive reasoning and rationalization impacting and influencing behavior. Advocating in silos, for demographics, groups, or organizations may bypass and dismiss the evaluation and analysis of operant mechanics and processes reinforcing and enabling harms, which are scaled across landscapes both physically and virtually. While developers build social platforms that capitalize on harms, through marketing while incentivizing behavior, it's been imperative to gain awareness of the underlying 'nuts and bolts' and manifestations of greed, envy, and ignorance impacting societies. Thus, every founder lays the foundation on which others operate therein. Intentions, motives, needs, aspirations, desires, and slue of other unseen forces, if unchecked, can manifest into generating benefits or harms.

The current operating model, as practiced in the United States, can reinforce victims or incentivize and promote them to engage with activities. Ignorance and lack of understanding can create mental voids in comprehension to weigh unseen functions and processes. Immediate effects, rewards, or awards provide the brain physiological, social, and fiscal stimuli which shortcuts judgment and reasoning by bypassing short and long term consequential evaluation. This concept is highlighted by the publication of Thinking Fast and Slow by Daniel Kahneman. Daniel identifies two mental mechanism that impact judgment, the 1st is emotional and impulsive; the 2nd is slow, deliberate, intentional and mindful. (Daniel Kahneman, 2011) Media podcasts and marketing can misguide and mislead persons through subliminal messaging towards affluence while creating false beliefs and mental models that subconsciously impact reason as to fit a false image or as many like to say 'Keep Up with the Jonse's' or in 2025, the Kardashians.

Marketers often rely on digital creation, now powered by Artificial Intelligence, to design and filter content to be more appealing, alluring, or enticing. Often, information is presented as lucrative, affluent, instant, and opportune as to impact behavior towards impulsivity. Subjects are consistently bombarded with stimuli, like advertising, oriented in triggering emotional states that surface familiar and familial feelings related to nostalgia, excitement, or even humor. Individuals are likely to accept presented choices and opinions without question, representing a psychological phenomenon known as Social Proof. Personhood empathetic traits, neurologically embedded mirror neurons, were incorporated by design for cooperation and collaboration as to build and flourish life. Due to exploitation of the human condition, many have been impacted by forming mental concepts based on what they've visually seen. (*What Is Social Proof? How to Harness Its Power for Marketing Success*, 2023) Individuals become primed, through repeated programming and stimuli regarding worldview, how to navigate in environments, and more critically how to perceive self and others. As such, this has founded and influenced skewed rationalization leading to compromised decision making that impacts optimal conduct and function.

Efforts in various countries keep sex work activities contained in areas. Benefits of prostitution can manifest as empowerment, in which subjects derive confidence from making personal choices about their bodies. According to the Human Rights Watch, "Criminalizing adult, voluntary, and consensual sex – including the commercial exchange of sexual services – is incompatible with the human right to personal autonomy and privacy. In short, a government should not be telling consenting adults who they can have sexual relations with and on what terms." (*Why Sex Work Should Be Decriminalized*, 2019) Summarizing, that individuals have free will to make decisions about their bodies, even if it's at their own determent, system dysfunction, and/or decay. Such advocacy highlights rationalization and the moral culpability that has affected policies and reinforced human and children trafficking, drug trade, and organ harvesting. Simply, advocates for self-determination misconstrueds the brains rationalization feature in categorizing what is actually an opinion as 'choice', rather than supporting conscious awareness as to evaluate choices, outcomes, and feedback for improved optimal functionality. It is recognized that human rights advocacy is intentional and scoped to support the development of personhood, however, within the context of unseen and seen variables and controls, self-determination manifests to self-compromising decision making. The lack of awareness in understanding that the body is a neurological extension of the nervous system and brain, may blind in evaluating the impacts conducted on the body and harms it transmits to neural system functions. Thus, allowing scalpel procedures or sensory abuse through body violations can have unseen impacts on moods, emotions, cognition which persons may not be able identify or root cause diagnose considering they are often opting for reinforcement 'quick fixes' or worse numbing legal narcotics. For example, the tongue is connected to 5 nerves within the brain; it would serve a higher purpose for persons to remain silent when internally turbulent, as practiced with mediation, to calm nervousness, anxiety, or other psych disorders. Thus, human rights advocacy produces paradoxical derivatives than originally intended to support personhood because they lead to self-compromise decision making leading to risk taking. Furthermore, these activities support industries that exacerbate exploitation of personhood like human trafficking.

The Human Rights Watch expands to infer that legalization allows the sex work industry to be regulated and provide access to resources like healthcare and condoms. Thus, a major benefit for legalization is having the ability to monitor and directly respond to and address the risks associated with prostitution. In accepting this model, justification to address consequences becomes a point in arguing for legalization measures, rather than providing such access regardless of established law. Moreover, this negates the impact of technology, like Search Engine Optimization, that present young adults with risk taking content as to frame justification in reasoning by presented rational arguments and mental models. For example, while adolescents conduct research for school, they tend to associate and align with group thinking as related to identity and may not understand the long term consequences of manifestations until they are made evident or visible. Teens may be exposed to sexualization content like those found in books, movies, or social media that are oriented towards pleasure seeking and hooking up. This reflects the overarching patterns of thought transmission impacting reasoning while aligned to democratic principles. This pattern of younger generations aligned with the democratic party and values, as represented by autonomy, is later shifted to republican endorsement due to deeper understanding regarding space operation and construct. Thus, to improve democratic societies and rule of law, establishment of value-based and evidence-based policies is imperative as to integrate optimal function, so as to serve all people.

Vulnerable persons in communities, like those in poverty, may develop a sense of insignificance in mental understanding of themselves and surface area, due to occupying eroded and deteriorated living conditions. Moreover, residents in fear of lowering their standard of living, while being bombarded with stimulus regarding 'have and have nots' or 'the elite' and 'working class' can be radicalized in reasoning and rationalization. This is further exacerbated with social programs stigmatization, like usage of 'food stamps' regarded with social status and worth by vulnerable persons like migrants and single parent households. The current operation may erode perceived self value and spirit collective connection due to earths fiscal economic function founded on digital infrastructure. Software companies, like Experian or Fico, provide 'ratings' regarding subjects fiscal scores with classifications as poor, fair, good, very good, and excellent. While the author of this book is writing this publication, her credit score is considered 'poor' considering she took a leap of faith to write what will uplift humanity; so trust such labels identifying worth couldn't be farther from the truth regarding soulhood and spirit connection. Although labels, classifications, tagging, and social scoring are there for efficiency for business function and national security; personhood value in soul and resonance cannot be quantified. Many recognize the properties of the soul including its value for being immortal and infinite. Whenever societies shift from this, persons are likely to misperceive or understand their true worth while mapping it instead of the the unseen to the seen. Until they're reminded... again... of how much they are love.

The unseen impact of fiscal economic operations on children, due to interpretation, may instill feelings of 'less than' while contributing to unseen forces and pressures that impact behaviors towards high achievement, bullying, prejudice, and out-casting. Feedback from community members reflected by the stories of orphans who grew up homeless or in poverty to later make it 'Big' show the impact of economic suffering on reasoning. When vulnerable groups have been conditioned on lack mindset and having unmet needs later become enticed towards marketing for 'affluence' and 'significance' by purchasing products and services for self-esteem and belonging. Moreover, judgment can be impacted through retained sense of obligation, as parents feel obliged to provide for their kindred a "better quality of life than they had". Scarcity and poverty mindsets can escalate and impact decision making when individuals are met with unethical or exploitation opportunities. Some may bite their tongue regarding known misconduct, accepting bribes or overlooking injustice, due to benefits, gains, or rewards received by complying and conforming with operating baselines. Such perceived benefits and rewards from exploitation or economic beneficial operations does not evaluate the long term generational or lifetime manifestations impacting each person, families, communities, and nations. Thus, monetary benefits or lack-there-of has had a wide range of influence on personhood reasoning and function.

While current judicial and correction system address the 'what' is happening factors regarding behavior, understanding the underlining reasons of 'why' they are occurring can provide a path for building better futures for all. The true measurement of worth has never been contingent on what can be quantified or qualified by people on earth, as seen with price tags and branding, as these are founded by men. Moreover, many of these misperceived affluent 'benefits', 'gains' and 'rewards' exploit insecurities, needs, desires, and identities or are founded on envy, greed, ignorance, abandonment, or negligence to later propagate in unseen ways by manifesting through people, places, and things. It would be valuable to re-frame worth and success, as presented by marketing, to scopes pertaining to character development and soulhood operation, reflected by the collective spirit of all people; as this will improve decision making between neighbors, businesses, communities, nations, and global regions. Thus, in the Human Race, success is not measure in having been first but rather by 'how many people we've been able to get across the finish line', so as to prevent the loss of life.

6

Global Response

Various countries have deployed varying models to address prostitution, known modernly as the sex work trade. For example, Sweden identified the 'Nordic Model', which makes buying sex illegal but prosecutes sellers, the sex worker (victim) in an effort to minimize the demand. (*Why Sex Work Should Be Decriminalized*, 2019) While countries have decriminalized prostitution and pornography for the purpose of monitoring and controlling outcomes and portray the illusion of choice and control, the current operation has expanded from the physical to digital spaces. Which are established on algorithms that operate on suggestion and reinforcement. For example, controlled area operations in the physical can uploaded digital content to the virtual and become accessible at scale to anyone with mobile devices. Personhood compromising websites, reflected by software infrastructure, have become personalized and tailored to groups presented by Rent.Men, OnlyFans, JustFor.Fans, and Pornhub. These operations have lured personhoods and contributed to human trafficking while aiding and abetting crimes that target younger populations, especially those vulnerable to suggestibility, influence, and power dynamic exploitation due to social economical needs, identity, or conditions. Even this author, while working in the service industry experiencing financial hardship, was approached by customers with a work opportunity that can provide better finances, than minimum wage, with stipulation that it would require camera recording. At the time the author didn't recognize this was human trafficking as she was operating in a place of work and engaging with her customers. But later upon reflection, it was evident that this was tactic exploiters use to procure victims in need.

Each person can support social good by improving tools, standardization, policies, and laws. As it stands today, with the complexity of psychological interpretation and rationalization, trust building in one infrastructure can be overgeneralized by persons to another. For example, teens exposed to the internet in controlled settings at school and social media at home, grow into adulthood having been operantly conditioned to use software, with trust, and later be introduced to other architecture that is exploitative due to multitude of reasons including but not limited to peer influence, financial needs, mirror and modeled behavior, and other subconscious influences based on norms, identity, needs, and interests. Thus, decriminalization policies have given way to reinforcement and enablement of harmful operations in the physical and virtual landscapes, due persons being legally allowed to build such infrastructure based on incentivization. A film called Sound of Freedom, by Angel Productions, raises awareness regarding the supply chain operation of human trafficking and expands to showcase the tactics and techniques used in exploiting insecurities and aspirations of vulnerable populations, including children. (Monteverde, 2023)

Presumably, within the United States, individuals and families operate in a deficit, in which economic returns from distributions of taxpayers money become consolidated to private companies via loans and grants. The minimal financial budgets afforded to individuals can provide better security from subjective high risk decision making, as represented with movie 'In Time' or stewarding fiscals funding responsibly due to energy transference and manifestation that impact populations. Maintaining a standard baseline salary range for national populations can provide national security, considering tools can monitor data and flag money laundering operations. Moreover, due to retirement accounts, everyone has a stake to maintain the status quo as to keep their retirement funding. Companies derive stakeholder buy-in with incentivization of funding returns while failing to disclose the means by which profit was retained. Such operations exploit the void created in perception and awareness while being continuously managed with public relation events and scripts. Technology companies that have utilized citizens' data to develop virtual infrastructure and tools, have used 'human resources' and exacerbated inequality by exploiting personhood energy for business and research ambitions.

Data tracking companies like Uber and Lyft have used employees' data to conduct surveying analysis and later invested in developing autonomous vehicles; where entities are worth billions whereas the common 50-year old employee is juggling home demands and child-bearing obligations, working two jobs, living paycheck to paycheck. Not to mention, scientists who spent decades on research, later have had their discoveries be used out of scope, intention, purpose or context. For example, discoveries identifying how psychology processes function and operate were utilized for propaganda and other exploitative means to manipulate or deceive populations, through color coding and branding. Such exploitations are consistently framed as lucrative opportunities while serving harmful purposes for self-interest outcomes.

While the privatization of business development is valid in expanding innovation to improve livelihoods; the inequality of sharing returns with those who have supported the work as having been a resource is not only unfair but generationally unjust. Countries that have used natural resources, were able to do so as a gift from God, which were to be steward within reason, intention, responsibility, and due care. Rather than exploitation for ambition, privatisation expansion or globalization, and energy reallocation in justification to grow infrastructure rather than build people up. Effectively, through marketing, life has been made about things rather than purpose. Moreover, this has contributed to societal and cultural shifts due to the over amplification of consumption based messaging centered around products and services. Instead of viewing life as valuable, people have viewed value only in that which is sold. And this has expanded to policies that sell people and soul essence. Marketing strategies for technology companies bolster the 'AI race' having subliminally shifted perception from valuing the 'human race'. While entities absorb natural resources, including people, to justify their means, such manipulation and deception does not go unnoticed.

Marketing aims to shift perception through subliminal programming. Such unseen forces aiming to make experiential matters about things rather than life have reckoned with higher forces. While a number of founders deny culpability and remain geared into their operational cycles, they express "We have money, they don't. They can't stop us." Such pride, arrogance, and selfish ambition blinds sight to weight interdependent operations in open spaces and negates the awareness that we occupy a magnetic energetic field. Acknowledgment and awareness of the likelihood that people use innovation or discoveries out of purpose or scope has been evident overtime and generations. Thus, the coming of changes manifested were brought by the lack of regard and observation of the ongoing evidence affecting populations and collective spirit.

Business infrastructure, digital and physical, has been used to derive, enable, and reinforce harms. Virtual data analytics and collections, operating on cookies, tag a Digital ID to a person while tracking their online whereabouts without the user's conscious knowledge or informed consent. The information collected about each person has been used to target and exploit individuals in various ways. Majority of sites harvested data for tracing and location tracking. As persons develop in education systems and into careers, they're assigned tasks without understanding the derivatives of their contributions and outcomes, both seen and unseen. Moreover, many are unaware that harmful operations are allowed in city zones that retain sales and provide tax returns. Meth hotels and trafficking zones are overlooked, while policing focuses on ticketing common persons for speeding with technology such as computer vision. Such imbalances to retain funding from harms are toxic to communities. While many think they can isolate such activities, technology like automobiles and mobiles phones have made such activities accessible by adults and children.

Not all tools, infrastructure, and laws are aligned to honor. Business operations and marketing has affected societies in exacerbating the culture of transactionism and consumerism for fame and affluence, while influencing personhood psychology to base experiences on tit-for-tat rather than generosity, selflessness, giving, or service. Thus, cultivating a culture based on gains and self-service convenience, may become misaligned to foundational values and principles. While each operates in silos, unable to weigh the objective causality in layers, each person's need to keep a roof over their head and provide for their children reinforces them to conform to established baselines of societal and cultural operations, maintaining a status quo that depletes energy and resources.

Personhood exploitation can vary by form like prostitution, porngraphy, advertising, click bait for continued purchasing, other subliminal or camouflaged incitation and temptations. Ongoing efforts to protect and guard persons from psychological, social, physical, virtual, spiritual, financial, intimate, and other soul harms manifest as employment opportunities. Majority of persons are operating within the scope of their identity, often forming beliefs founded from the internet, in socioeconomic classes, careers, familial beliefs, and other mental frameworks and programming as to be able to distinguish and objectively weigh variables, outcomes, and manifestations. Generations are left to receive advice from elders that may seem outdated or irrelevant due to technological development and infrastructure setting changes. This can furthermore develop divides or disconnection between children and parents, industry workers and peers, and countries of varying economic levels producing division, misunderstanding, or conflict. Media, especially social platforms, have clustered people into subcultures that amplify and reinforce false narratives, falsehood, discord, and false beliefs impacting behavior based on justifications and inferred misinterpretation. This is likely the unseen cause of the January 6 incident on the United State Capitol, which reflects national security threat and risk. The same tools that were created to keep persons protected are now being leveraged to amplify harm, discord, and incivility.

Instead of people mastering things, things are mastering them by tugging and suggesting to them what to do, who to talk to, where to go, who to be, what to wear, and more dangerously how to think and perceive. Marketing and cameras shift perception from looking out into the open to specifically in a direction with distortions and illusions as to sexualize, influence for affluence, status, or power mode operations. In effect, this calibrates energy and focus towards others, shifting self-concept, and worldview by established narratives that are not founded on true love but other motives. Millennials grew up with mobile phones that primarily had one camera on the outside of the case, to later be facing them amplifying 'selfies', and now geared into facial recognition software that can detect their emotions and impact hiring. Such changes in technology have impacted self-interest and self-centeredness and 'put yourself first' movements while categorizing such behaviors as 'narcissism' and blaming soul function on parental conditioning. Modifications and tweaks, although subtle, can affect behavior and lead to paradoxical manifestations in founding social media for connection leading to a loneliness epidemic. Friends have belittled and looked down upon others who have earned less, parents become disdained with the accomplishments of their children, and societies erosion of truth and trust have been replaced with popularity, engagement, and ratings.

To truly address manifested atrociousness, shifts in cultivating spaces founded on gratitude, acceptance, kindness, presence and care taking can equip and improve mental functions, clarity, understanding, and awareness. Overall, the global impact of allowing the exploitation of personhood, on any level, whether it be emotional, social, psychological, spiritual, financial, relational, personal, and/or intimate ends up costing humanity a lot more than monetary wealth. Exploitation erodes trust and connection while exacerbating illness, disease, and conflict. Furthermore, scaled innovation should not be undermined for its impact. Global strategies ought to align to value and evidence-based conscious policies that uphold the dignity and wellbeing of all people. Informational sources require to be evaluated as to identify if they incorporate false beliefs, discord, or foul frameworks. To improve consciousness, current operations in silos and privatization need thorough evaluation as to eradicate the root causes contributing to harms like those derived from envy, greed, or toxicity. Rather than applying risk models that accept harms, persons are required to identify strategies to eradicate atrociousness.

In the United States, most perceive price tags while wagering soul essence, innocence, and dignity. While doctors bear having to pay hundreds of thousands of dollars in student loans, balancing mortgages, and other financial demands, some professionals may resort to violating or committing harms like fraud due to pressures or duress as to pay bills. Other times, marketing and affluence can develop motivations that drive persons to violations in an effort to feel significance which may be manifested from having developed in impoverished communities. Exploitation of persons is rather at times not reflective of personhood malicious intent but rather manifestation as to sooth or meet psychological, physical, social or financial needs. This can expand to reflect white collar crimes with acceptances of bribes, kickbacks, and other morally corruptible acts that manifest from rationalization, due to fear, blackmail, or coercion. This is reinforced by messaging production, through marketing, which amplifies fortune seeking, fame, and allure.

While it is imperative to meet needs, a large number of spiritual and religious guidance has echoed the same sentiment regarding accumulation as it relates to fasting as to not corrupt cognitive function. Spiritual awareness and clarity may not be made manifest in consciousness while persons are operating in cycles of production, consumption, and accumulation. Thus, exploitation of personhood can be compared to cannibalism, eating flesh, due to the nature of entities/corporations deriving their sustenance from consuming resources, represented by the human resource departments each has. Such consumption and exploitation of resources can represent irrational wants, desires, ambitions manifested from psychological, cultural, and personal justifications. Countries when given the opportunities to send tractors and build civilization resorted to sending weapons. Furthermore, corporations have derived capital based on misguiding populations and reinforcing cycles.

The environment was meant to be maintained and stewarded with respect and care, including people. Even with all this shared for improving awareness as to identify and distinguish moral corruption manifestations, recognizing who and what to blame is less meaningful as to identify pathways to cultivate change. Ultimately, the rule of law, even as it pertains to prostitution, may negate the domino effect, consequences, and escalations. Small technological developments have had a profound impact on personhood operation and function. Further, a large number of business development become manifests as to address consequences and outcomes. While many may advocate for choice, being able to perceive and recognize the thoroughness of operation in landscapes has become about money rather than integrity. Experiential outcomes thus have become a reflection of biblical expressed prophecies, based on choices contradictory with higher guidance.

The supremacy of law shall be founded on truth, infused with layered analysis of variables impacting processes, systems and consciousness. While the soul has been designed for sovereignty, current operations and functions have tethered and produced dependency and attachment. For example, persons are now tethered to digital mediums, in an effort to receive services or be able to buy products due to merchants integrated transactional digital processing systems that are supporting data collection for national security monitoring, marketing, and even hiring based on sentiment analysis and computer vision technology embedded within programs and software. So, it isn't shocking that we really have done this to ourselves by accepting presented options, norms, and status quo while aligning to willful blindness due to resistance to change and accountability while ignoring surface occurences. Furthermore, many of those who dared to speak 'truth' have been persecuted or prosecuted, shunned, and/or deported or exiled. Ultimately through sharing information with intentionality of cultivating awareness, perhaps we're at the pinnacle of supporting the cultivation of character and personhood while paving a path for a future that is safe, nourishing, just for all; should we not continue to do as we've always done but perceive, acknowledge, and remediate what has been surfacing.

7

Evidence

The research findings are presented in charts, graphs, and tables with brief explanations aside exhibit numbers. Data has been generated from agency resources, including but not limited to the FBI Uniform Crime Reports (UCR), National Institute of Mental Health, San Francisco Government Data, SF 2024 Financial Annual Report, and on-foot field analysis and surveying. Each section represents a domain interdependent and influential in the response and reaction process within the setting. Data collection and research has been completed from 2023 through 2025. Lead researcher, aligned to faith, has a background in criminal justice, cybersecurity, artificial intelligence, data science, health sciences including biochemistry, psychology, sociology, customer service, and business.

I. Public Safety

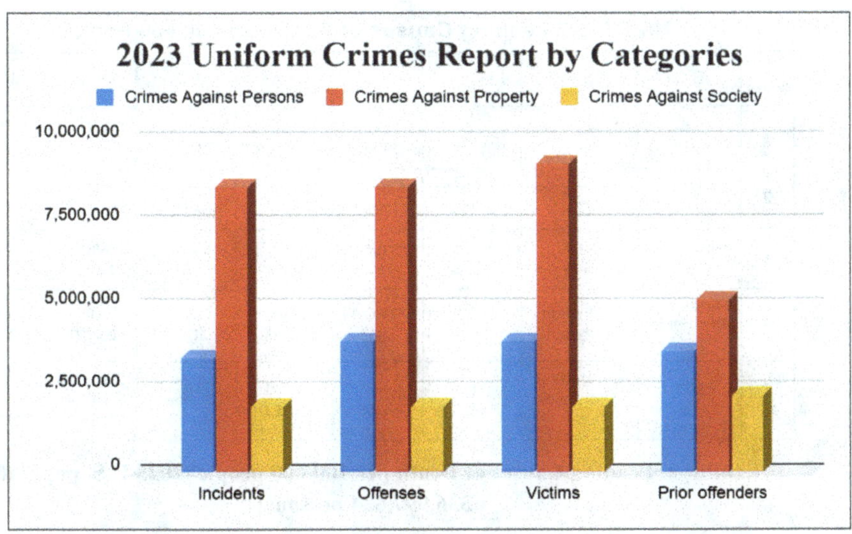

Exhibit 1. 2023 National FBI UCR Case Count by Categories, accounting to 12,342,260 cases nationwide.

**Blue, Against Person: Assault, Homicide, Human Trafficking, Kidnapping and Abduction, and Sex offenses.*

**Red, Against Property: Arson, Bibery, Burglary, Forgery, Vandalism, Embezzlement, Yellow, Extortion, Fraud, Larceny, Motor Vehicle Theft, Robbery, and Stolen Property.*

**Orange, Against Society: Animal Cruelty, Drug Offenses, Gambling, Obscene Material, Prostitution, and Weapon Law Violations.*

Graph represents 2023 crime incidents categories against a person, property, and society. Crimes against property represent the highest number of crimes committed. While crimes against person and society are significantly less represented. Speculatively, acts committed against a person are less likely to be reported and/or prosecuted due to the complexity interwoven with domestic violence, workplace violence, and psychological manifestations stemming from emotions and passions, needs, or influence.

II. Health

2021 Top 5 Leading Causes of Death per 100,000 People								
Rank	5-9	10-14	15-24	25-34	35-44	45-54	55-64	All Ages
1	UI 827	UI 915	UI 15,792	UI 34,452	UI 36,444	COVID-19 36,881	MN 108,023	HD 695,547
2	MN 347	Suicide 598	Homicide 6635	Suicide 8,862	COVID-19 16,006	HD 34,535	HD 89,342	MN 605,213
3	Homicide 188	MN 449	Suicide 6,528	Homicide 7,571	HD 12,754	MN 33,657	COVID-19 73,725	COVID-19 416,893
4	Congenital Anomalies 171	Homicide 298	COVID-19 1,401	COVID-19 6,133	MN 11,194	UI 31,407	UI 33,471	UI 224,935
5	HD 66	Congenital Anomalies 179	MN 1,323	HD 4,155	Suicide 7,862	Liver Disease 10,501	Diabetes Mellitus 18,603	Cerebrovascular 162,890

Exhibit 2. Leading Causes of Death per 100,000 people (2025 U.S. population 336,997,624 persons)

**Unintentional Injuries (UI): Includes Opioid (Drugs) Overdoses, Motor Vehicle Crashes, and Unintentional Falls.*

**Suicide*

**Liver Disease*

**Diabetes Mellitus*

**COVID-19*

**Cerebrovascular: Strokes*

**Heart Disease (HD)*

**Homicide*

**Malignant Neoplasms (MN): Lung Cancer*

**Congenital Anomalies: Birth Defects, Down Syndrome, Defects, etc.*

Table represents leading causes of death attributed from consumption and activity. Operations are interwoven through infrastructure, poducts, services, and peer associations influencing and impacting decision-making. While persons are responsible for their choices, it is imperative to weigh development, mental models, conditions, states as to trace comprehensively root causes.

III. Business

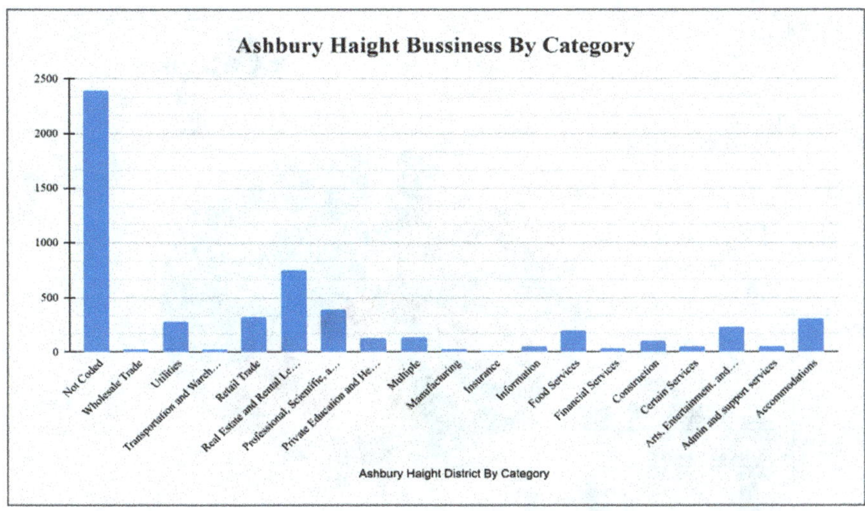

Exhibit 3. Ashbury Haight Business by NAICS Code Categories

Graph represents the 2024 San Francisco registered businesses inventory by the SF government, within the Ashubury Haight District. 2387 of 5416 inventoried business are 'not coded' with an affiliated industry.

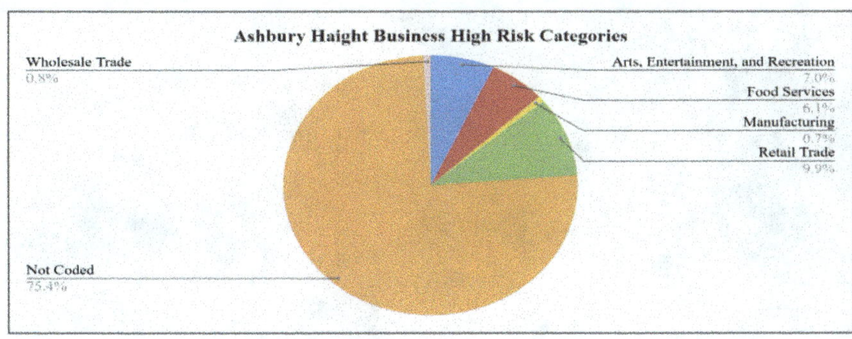

Exhibit 4. Ashbury Haight Business Categories providing Risk to Locals

Graph represents district industry categories allocating spaces, products, and/or services with risky offerings. This includes not coded NAICS categories, representing 58% of all operating business.

IV. Government

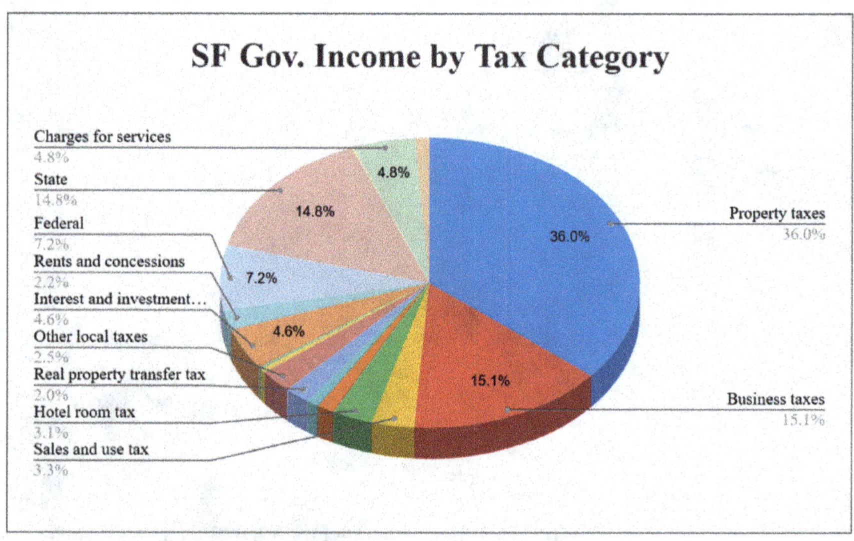

Exhibit 5. 2024 San Francisco Government Income by Tax Category

Business, sales, and property taxes account to 54.4% of the SF government income. This current government income is reflective of high risk business operations attributed from harmful substance and paraphernalia sale.

V. Social Psych Analysis

Exhibit 6 Subliminal Messaging

Subliminal messaging, like advertisement and marketing, operates on the subconscious and conscious levels of sense processing. Such stimuli is able to entice, prompt, suggest, and/or influence within the ecosystem. Current operations produce, seen and unseen, influences via physical and virtual stimuli triggering thoughts impacting judgment and reasoning thus influencing behavior and habits. Consumption through the senses, overtime, may and can enable the propagation of cyclic habits, as residents move through open space accessing products and services presented.

VII. Malicious Products for Sale

Name	Origin	Type	Packaging	Psychotropic Effect	Developer Education	IRS EIN	Legality
Ripple+	London, UK	Smoking	Electric Pod	Irritation of the throat and lungs, coughing, and other respiratory symptoms.	BA Economics, Master Degree	Not Found	Requires Testing
Professor Seagull Salvia Divinorum, SF SMARTSHOP LLC	Vilnius, Lithuania	Ingestion	Herb	Confusion, paranoia, hallucinations, and delusions	Not Available	Not Found	Illegal
ARTET GROUP INVESTORS, LLC	Brooklyn, NY	Ingestion	Delta-9 THC Drink	Anxiety, dry mouth, increased appetite, memory loss, red eyes, slowed reaction times, rapid heart rate, confusion, dizzyness, hallucinations, paranoia and panic attacks, nausea and vomiting.	BA Communication	Not Found	Illegal
DMT the Spiritual Molecule	Los Angeles, CA	Information al	Book	Thought Inception	Dr Psychiatry, Pineal Gland	Not Applicable	Illegal
Magic Shrooms Gummies Golden Missle	PharmLabs San Diego, Magic Brand Holdings, LLC	Ingestion	Mushrooms	Paranoia, hallucinations, and delusions	MS Statistics, BS Biochemistry	Not Found	Illegal
Three Spirits Drinks, Beyond Alcohol, Inc	Delware, NY	Ingestion	Plant' Caffiene Mocktails	Anxiety, Dehydration, Dizziness, Fast Heart Rate, Headaches, Insomnia, Restlessness, Alertness, and shakiness	Geology, Public Health,	Not Found	Requires Testing
Rapé (Hapé) Shammanic Supply	New Mexico,	Inhalation	Smokeless Graded Tabacco, Tree Ash, DMT	Dizziness, nausea, sweating, vomiting, and diarrhea	International Association of Kambo Practitioners	Not Found	Illegal

Exhibit 7 Available Sale of Disguised Control Substances

Controlled substances disguised as 'medicinal' or having 'plant-based' properties, available for sale within the Ashbury Haight District. Product brands are not medically verified nor found to be IRS registered. Product ingredients produces effects that are harmful to consumers, amplifying anxiety, paranoia, and confusion. Moreover, sales present immediate effects without representation of short-term and long-term effects and consequences of use on health, function, and consciousnesses. Such substances are introduced and prompted in high trafficked settings like festivals, night clubs, and parties.

VIII. Field Analysis

Exhibit 8. HIV Prevention Billboard Signage is Signaling to Demographics and Nudging Towards Risk Taking Activity.

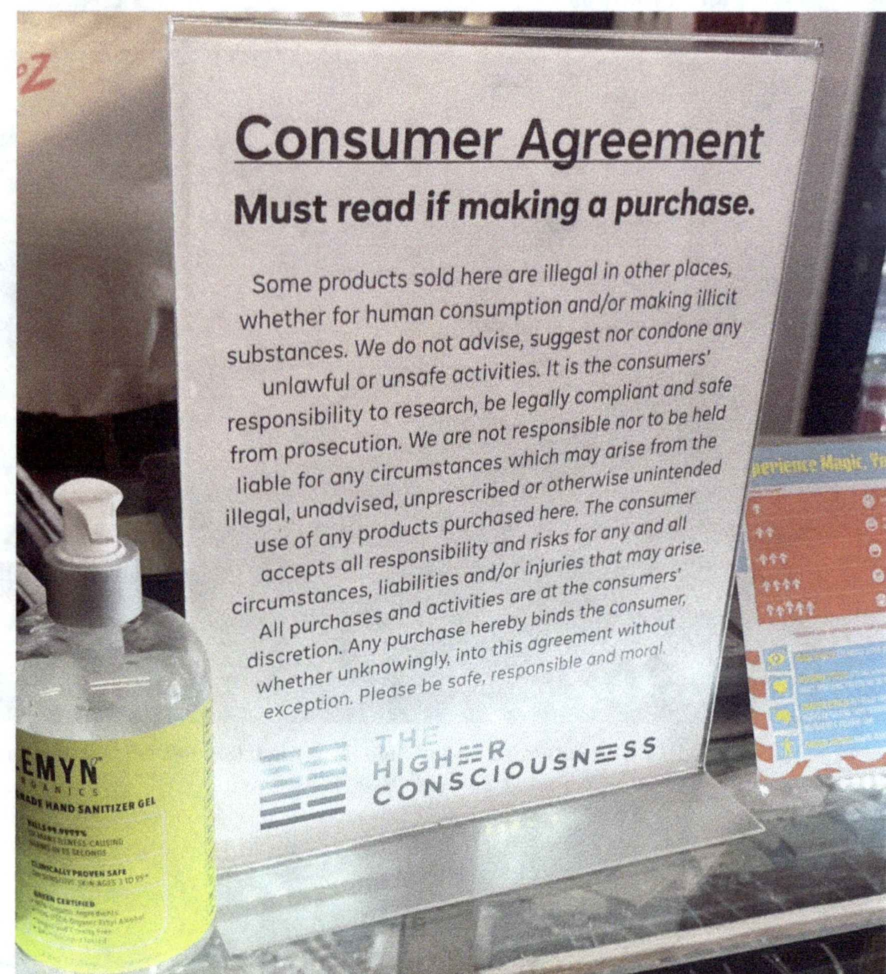

Exhibit 9. Consumer Sale Agreement Presented at Time of Sale. Ignores operational speeds by consumers in public spaces or distractions in group conformity or peer pressure.

Exhibit 10. Psychotropic Controlled Substance Business Misrepresented as 'Higher Consciousness' when Scientifically Reflects Brainwave and State Inhibitions.

Exhibit 11. Drug Test Kits Featured Outside Psychotropic Business Store, Reinforce Drug Use Culture.

Exhibit 12. Inhalation Substances Dispenser, Providing 'Beta Inhibition' as to Effect Heart and Brain Rhythms.

Exhibit 13. Psychotropic Substances, Represented as 'Plant Medicine'

Exhibit 14. Sight Seeing Tour Bus in Critical Risk District by Patrons of All Age Groups.

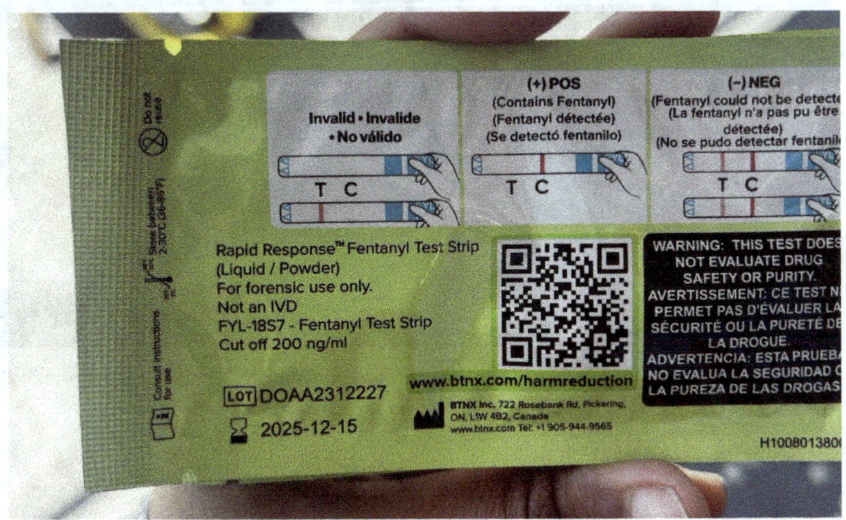

Exhibit 15. Fentanyl Test Kit, Reinforcing Drug Use Culture

Exhibit 16. Smoke Shop Store Front, Providing Access to Paraphernalia to Walking Pedestrians by All Age Groups

Exhibit 17. Inside Smoke Shop Business, with Paraphernalia Access

Exhibit 18. Smoke Shop, Presenting Paraphernalia, Shaped as Weapon and Representing the 'Unseen Weapon' contributing to the "Seen War'

Exhibit 19. Needles Found Outside a Local Business

It is recognized that customers, patrons, visitors, residents and/or tourist of the Haight District are not likely weighing the subconscious influence of designs, messaging, or product offerings while navigating, purchasing, or consuming due to familial or peer engagement, distractions, cultural bypassing, historic district 'theme' normalization, or availability of offerings. Many of the presented materials and contents in the area incorporate skulls, objects with malshaping or messaging, including derogatory designs or obscenity. To improve understanding, it is imperative to distinguish the details contributing to personhood attire marking and labeling, while cultivating awareness and mindfulness of intentional product production and consumption. Many of available ingestion includes offering that subdue consciousness via availability of psychotropic inhibitors and substances as well as paraphernalia or dispensers for administering compounds.

Large number of the population in this district are adolescent, as there includes coffee shops, vintage clothing stores, arts and crafts shops, food and restaurants, cyclist merchants and more. The main takeaway from these exhibits reflect the traced foot pathways traveled by diverse demographics, which influence thoughts, actions, behaviors, routines, and generational cycles that could expand or manifest in other spaces due to innovation and mobility. California, especially San Francisco is known around the world for tourist attractions like Golden Gate Park, Golden Gate Bridge, Big Tech, and International Business Operations. Moreover, many of the locals who were locally surveyed report being born and raised in the area and have operated on survivable wages without feeling inspired to leave due to the mediterranean like climate. Higher guidance provides subjects the opportunity to build better, but the district is considered historic reflecting hoarding operations and peronhood subduing due to sensory abuse offerings, including social events.

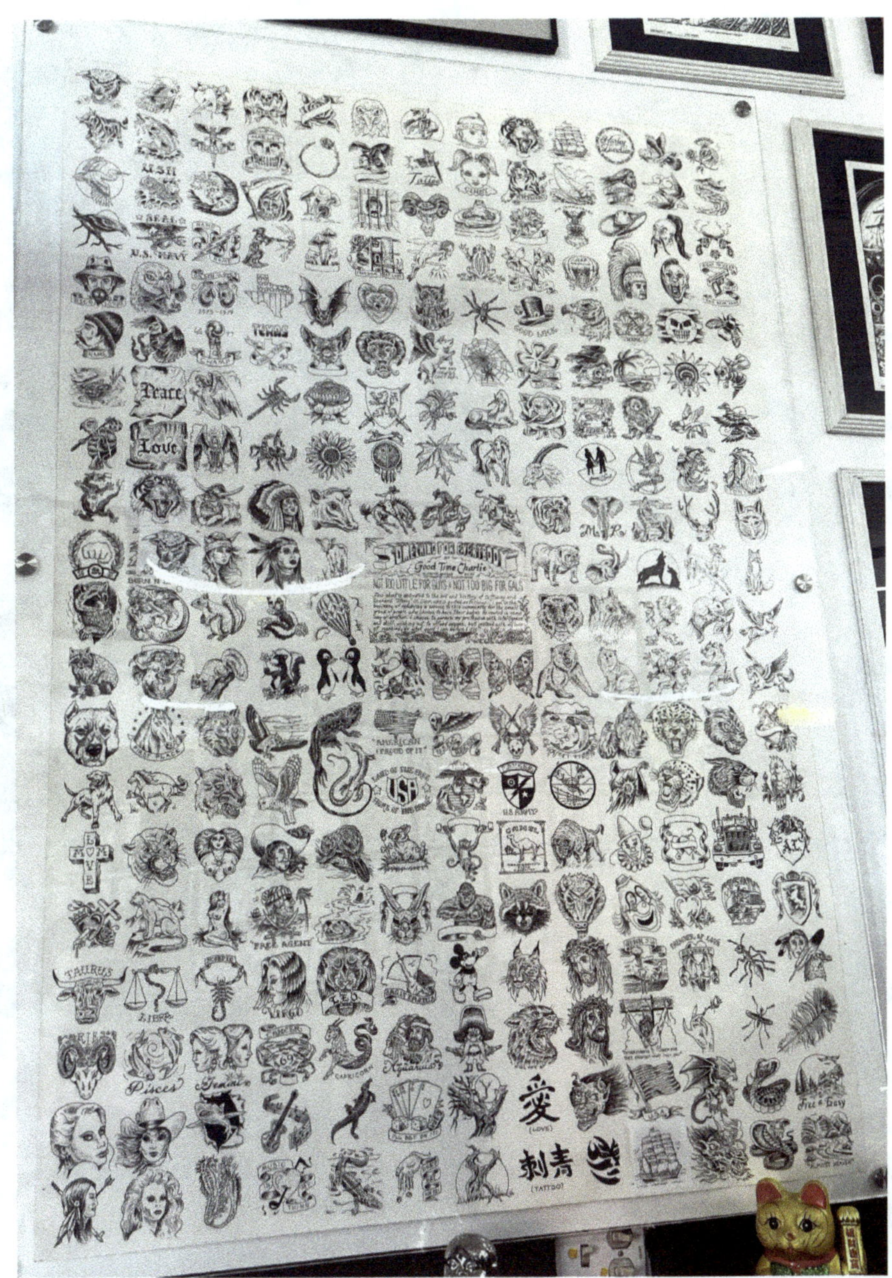

Exhibit 20. Available Tattoo Designs in Tattoo Parlor

Exhibit 21. District Art

Exhibit 22. Derogatory Messaging on Clothing

Exhibit 23. Sock Designs

Exhibit 24. Accessories Store Showcasing Skull Rings

8

Change Mandates

Technology

It has been made evident that innovation in the context of devices, projecting images, have had a wide ranging effect on perception which has distracted, distorted, misguided, and/or misled persons in thoughts and beliefs leading to manifestations of behaviors, activities, habits, routines reflected with societal patterns. Furthermore, applications within mobile devices have amplified divides, discord, and reactive impulsivity leading to conflict and violence, health disorders, erosion of trust, risk to national security and many other outcomes. For this reason, the main proposal is to terminate and discontinue image projecting device manufacturing contributing to seen and/or unseen harms. While this phase-out process is being completed, new activities are phasing-in to integrate positive synergy, harmony, cohesion, and nourishment. Listed are change mandates to be implemented by industries. It is imperative for leaders to be aware of conflict of interest conditions, binding agreements, or bondage interwoven into roles based on justification, rationality, selfish motives and malintentions. For this reason, this publication relays insights from higher guidance and intelligence that has traced root causes contributing to manifestation impacting thoughts leading to harmful outcomes, seen and unseen.

Webcrawlers are bots, software programs, equipped with artificial intelligence and machine learning capabilities that scan and search the internet with specific defined instructions and conditions, like those conducting website compliance audits. Webcrawlers support search engines to optimize sites for user experience. For this reason, webcrawlers can be utilized to support the eradication of harmful online content and media that depicts violence, abuse, or such that reinforce falsehood or malinformation enticing towards harms. (*What Is a Webcrawler?*, n.d.)

Synthetic ID would be embedded in the meta data of artificially generated content as files do with digital hashes or signatures. This can improve digital content integrity, brand trust, and user confidence. This can illuminate a visible graphic user interface tag, identifying AI generated content. Enabling internet users, of all abilities and demographics, to distinguish artificially generated or digitally engineered content, opinions, and/or digital abuse material. Tags can be made visible, clear, and recognizable through the space time continuum by generations, across all nations, regardless of native language or origin. It is imperative digital content is labeled correctly for accuracy and validity as to support the filtration of false, illegitimate, distorted, or misleading material impacting subcultures, diverse populations, and vulnerable groups. This will strengthen societal cohesion, national security, synergy, productivity, culture and morale.

Search Engine Optimization (SEO) orders websites relative to user search inputs entered within the search prompt field, as reflected with Bing or Google. For example, when a user searches 'apple', search engines will optimize the results to the highest trafficked websites based on engagement, rating, and/or relevance. Thus, current top search results for 'apple' would place

'Apple Inc Technology Company' atop results, rather than an actual apple fruit. Such features have been under scrutiny, due to skewing search results towards major commerce retailers like Amazon, Walmart, Ali Baba or Ebay rather than small local businesses, unless it pertains to conditions like maintenance or emergency response. Furthermore, SEO has had a large impact on internet users' mental framing of personhood beliefs and identity. Results map to websites based on relevance, which trace to organizations or entities that promote ideologies impacting behavior, activities, and routines. In the worst cases, internet users develop false beliefs with altered perception regarding the experience and act therein with associations rather than awareness or consciousness. New technology like chatbots, such as Copilot, have incorporated search engine optimization to ensure search results are presented in a pane immediately visible atop for internet users. While data is being collected regarding user search prompts with user IP and device ID, presented information within the search pane may not be credible, safe, or valid. SEO features inhibit critical thinking and deter due diligence in evaluating hosting site and source information, since chatbots pull data through machine learning and AI capabilities.

In an effort to improve and establish digital order, categorizing websites with color coded tags that identify opinions, private corporate material, health organizations, government agencies, and even independent journalism would organize the internet library. While SEO organizes website records based on 'most read' results rather than alphabetical order, it's imperative to assess the current digital operation and its impact on generations. This is to ensure vulnerable populations, the uninformed, or disabled don't get left behind, digitally abused, or misled.

Domain Registration standardizations are required to trace and hold accountable website hosts who upload violating or abusive content, malproducts, and malservices. Moreover, domain registrants need to integrate education curriculum pertaining to accessibility, societal impacts, ethics and integrity, internet use and abuse prior to obtaining domain to host a website. It is imperative to cultivate safe spaces both physically and digitally through digital accreditation of website registrants, while upholding integrity in the development, deployment, and publishing process. This is to include social responsibility as to prevent human trafficking, drug trade, organ harvesting, and other exploitation operations facilitated through internet mediums. Such learning can inhibit digital abuse from scaling to populations, improving digital publishing processes, digital trust and national security. Further, websites shall be crawled to assess if registration information regarding website content is explicit, violent, obscene, indecent, etc. And if it is congruent to registration information, categories, and codes.

Firewalls are devices that control network traffic with predefined conditions and rules to secure digital infrastructure. (*What Is a Firewall*) Many nations, unlike the U.S., have incorporated higher internet controls, including firewalls by blacklisting or blocking harmful sites from operation that are impacting societies or persons negatively. Utilizing technologies like firewalls can cultivate safe virtual spaces and significantly reduce harmful web traffic operations that mislead, misguide, prompt or tug end user behavior towards self-compromising activity. While content moderation filters harmful content on the application layer, firewalls prevent malicious traffic on the network layer of the internet. The majority of corporations, governments, and enterprises incorporate firewalls by hiring cyber professionals to manage firewalls and domain activity. However, internet users have predefined firewall default settings that they may never check or are informed to modify to incorporate better protection or controls. For this reason, digital literacy gaps experienced by migrants, parents, adolescents, and other social groups like the disabled need guardianship implemented by design. This is not to suppress expression or impede amendment rights, but rather support the cultivation of safe spaces that allow for social responsible function and expression. Firewalls can cooperate across national and regional boundaries to support the development, safety, and health of all people through uniform standardization.

Firewall rules shall never be used by means to abuse, suppress, or oppress persons with dissenting opinions, as each person represents conscious manifestations operating in seen or unseen ways. For this reason, pathways to sharing feedback by varying age groups and diverse populations are to be valued, within reason, responsibility, and intent. Due to potential abuses of power and corruption, any information blacklisted or shadow banned shall be investigated for merit, validity, and legitimacy as to ensure actions are not in scope of retaliation. This is to ensure natural cooperation to develop societies authentically with integrity. This ensures the network is founded on transparency, fairness and justice by design for all.

Product Design will require evaluation in scopes of influence and impact on awareness and conscience. For example, due to mobile device camera designs being front facing, users became preoccupied with the self by taking 'selfie' portrait photography. This cultivated a culture that has operated on 'fitting in', amplifying envy, jealousy, and slue of other cognitive inferences and operations impacting health and morale. Instead of people building connections between each other, now they're preoccupied looking the part online, cycling through experiences to fit in, focused on promoting materials, and slue of engagement role play reliant on uploading digital content and materials. Device mediums can impede true connection given the cognitive influence digital mediums have on sense making, like attention.

Devices were primarily immobile, allowing individuals to roam freely. However, currently as of 2025, devices have become accessories and tethered to the palm of users hands with contingencies and dependencies for use in scope of retail, commerce, and travel. Thus, while product designs are inspired for user functionality and efficiencies, it is imperative to assess product designs impact on health, morale, and society. Furthermore, it would be beneficial to evaluate user feedback, including litigation, public comment or sentiment, and other pertinent facts regarding digital use or abuse.

Litigation and case precedents processed by the tech industry require intensive review and appropriate response to ensure surfaced matters are not in actuality reflective of product complaints or harms that have escalated to a degree requiring public oversight, mediation, and attention. For example, when a business is providing poisonous food to a customer who later becomes sick and exhibiting symptoms; the consumer may not be able to share their impairment or injury but professionals or peers evaluating or witnessing the matter can attest to those facts and alert regarding harmful misconduct or business maloperation. While many cases require victim presence to submit legal matters, some who have lost their life, in fear, or operating with religious alignment may not assert their claims of wrongdoing. Furthermore, material evidence regarding harms may go unreported or buried with absent persons. For this reason, evaluating cases in civil or criminal scopes shall be conducted by technology integrity officers as to address root cause manifestations and prevent culpability occurrences or misconduct omissions.

Cooperation and collaboration between industries through conferences like technology, medicine, social work, and business can improve problem solving in addressing societal harms. Plea deal processes tend to brush off, bypass, and overlook root causes contributing to harms. While individuals operate in subjectivity, they may be unable to weigh causal relationships of product deployment, influence, and impacts. Moreover, outcomes may be overlooked due to complexities of roles, responsibilities, culture, norms, unseen forces and variables. To support the overall function and improvement of civilization, it would be imperative to view cases as conscience surfaced signals representing symptoms experienced by the general body, queuing warnings regarding effects and impacts. Methods in addressing causal manifestations can be in scope of applying 'processes of elimination' in which risk managers or business owners can remove or undo products or features known to have a harmful impact. This will ensure safety is prioritized above company profit or branding, including assessing the seen and unseen costs to civilization and generations life cycles.

Compliance and standardization shall be scoped to include civilian experience feedback, seen and unseen metrics that include cost benefit analysis beyond company objectives, rights violations, and life lost. For example, risks shall include unseen metrics like technology impact on user productivity, student development test scores due to social platform reel distraction, user time wasted, health cases attributed from risk taking activity suggested through marketed reels and algorithms, and end user product quantitative survey scores. This is to support transparency, national security, and ensure comprehensive policies are founded on awareness and materialized facts.

Virtual Reality is a highly immersive digital experience that stimulates the optical sensory process of a product user. This is often coupled with auditory cues to provide a multi-modal or multiple modality sensory experience. VR is highly digitally simulated and founded on software code with a specified intent and purpose like that seen in gaming. VR system capacities and capabilities are computationally fast, making virtual reality appear to feel and look 'real'. This can manipulate end user awareness due to graphics enhancements projected via artificial designs given the quality of hardware and software components. While VR provides a better seamless experience for end users, young adults should err on the side of caution while using VR technology as to ensure their sense processing is not distorted due to use or overuse. Moreover, conscious awareness may be influenced from virtual reality utilization that seep into and impact personhood worldview, self-concept, and life philosophy. As innovation continues to grow, so will the need to ensure that technology does not impact users negatively or influence cognition in ways that manipulate, distort, misguide or comatose.

Marketing

Standardization of the marketing industry is imperative to prevent distortions and manipulation of perception regarding products, services, and experiences. Alignment to values of social responsibility, accountability, authenticity, and transparency will improve marketing messaging and yield benefits to improve societal integrity, trust, and safety. Moreover, business owners will be held to higher levels of standards in assessing the impact of their operations beyond benefit cost analysis correlated to capital gain margins or performance. For example, it would benefit brand development and brand trust to assess service and product effects, short and long term impacts and consequences on personhood function and society. Moreover, business owners will need to complete learning in social responsibility, wellness, ethics, integrity, and cultural influence and impact when renewing licenses and permits.

Target audience scopes will need to include empirical evidence testing prior to public deployment or publishing. This includes but is not limited to authenticating and evaluating messaging for bias, falsification, deception, and/or manipulation of material fact regarding product or service properties, contents, ingredients, performance, and other guarantees. Furthermore, target evaluation may include but is not limited to unseen and seen forces, intent, purpose, procedural processes and protocols. This is to ensure presented information is not targeting or triggering impulse, emotions, urgency, or other subconscious stimulus influencing or priming behavior.

Advertising messaging shall be limited, reflecting psychological stimuli serving sizes. For example, each device/IP shall have a set limit of ads played per day with alternating categories of industry content to prevent over saturation, purchase cycle reinforcement, and monopolization. Perception and priming towards material consumption like clothing, cars, psychotropics substances, travel, and indulgence shall be extensively moderated and regulated. This is to support that the life experience and value of life is not based on material things and material gain. Incorporating psychological stimuli serving sizes will cultivate psychological safety, improve conscious consumption, conscious production, societal morale, productivity, synergy and harmony.

Advertisement presentations shall reflect product or service specifications and ingredients information as well as product development process, effects, including short and long term consequences, and design explanation. This is to eliminate manipulation, deception, or misrepresentation that triggers emotions, impulsivity, grandiosity, over indulgence or greed. Furthermore, this will improve transparency and ensure infused energy in production operation is reflective of acts and intents aligned and interwoven with integrity in every business decision and process.

Research and Development

Projects, programs, and initiatives that manipulate, deceive, entrap, or entice towards harm or risk-taking are to be rejected, never funded, nor approved for permits or sponsorship regardless of justifications or rationalizations; especially those under the disguises of 'national security', 'sting operations', 'research and development', or 'religious purposes'. No person shall be studied or experimented on, in any capacity, without authenticated consent to be acquired every year with formal processing procedures and conscious informed consent by participants, including witness notarization or presence of trusted source like a family member; taken yearly. Furthermore, no such consent shall ever be implied or inferred on nonverbal language, but be made physically through coherent written and verbally methods. Study or experimentation permissions shall never be taken under unconscious, unaware, or inhibited mental states. Study spaces shall always provide signage and written documentation of experiment and study scopes, purposes, objectives, protocols, procedures, terms conditions, and any other pertinent information. All studies shall be completed in designated areas, reasonably known to any person and general public for their purposes. There shall be visible, accessible and clearly written signs, banners, and postings regarding study or experimentation purposes, processes, procedures, and intended objectives.

All research and development initiatives shall be prioritized in lower order than those upholding safety, security, and preservation of life. Any action that misleads or frauds the public morale or health will be held as a violation to life, subject to imprisonment. Monetary fines of guilt shall be within means and obligations of the subject, as to not add undue duress on household or childbearing responsibilities. There shall be no covert experimentation whatsoever under any condition, especially as it pertains to national security, due to the nature of the human condition. Studies advertising promotion and incentive rewards or awards shall be canceled immediately from propagating in media or through digital mediums; especially those reaching children, adolescents, and young adults. Studies shall only be obtained, for participation by clients or patients, through designated accredited practitioners settings like doctor offices. Furthermore, consent and agreements shall be reviewed by third parties like witnesses, legal correspondents, and health practitioners; with evaluation of effects, short and long term consequences. This is to ensure covert, rogue, misrepresented, ignorant, misguided or disguised maloperations, malproducts, or malservices aren't facilitated through back doors or mediums.

Professionals conducting studies or experiments on subjects shall be clearly marked with designated tags identifying legal name, role, department, badge number of agency, and affiliation(s). No person accessing healthcare services is to be experimented on, studied, or monitored for scientific discovery or advancement without written and verbal consent by the patient or client. Any data obtained from covert operations shall be destroyed and no longer retained, especially in cases in which subjects have died, unless within the scope of judicial procedures set for justice. The application processes for studies shall be lengthy, spanning months and years before approval, with conditions requiring thorough medical, social, psychological, nutritional, field landscape surveying regarding product offerings to identify root causes. Relocation or setting changes shall be weighed in improving optimal health outcomes.

Sun rooms and natural green rich spaces shall be developed as extensions to health care facilities for improving healing and providing calm, stress free quarters. Any subjects exhibiting visible incoherence or signs of incoherence during or after studies or experimentation shall not continue to be studied or experimented on, as to weigh if experimentation is causing disruptions, disorders, and/or disturbances. Nutrition intake inventories shall be completed as to weed out toxic compounds or chemicals ingestion. In the event any surface area, represented by a zip code, presents an increased number of ailments, disorders, or illness, a thorough environment investigation and field analysis shall be completed as to identify and determine any root causes contributing to cases, including but not limited to toxic distributions, obscene or violating activities, malservices, or malproducts offerings. Research and development integrity officers shall evaluate common age group habits, involvements, activities, and consumption to ensure variables causing common conditions, for participation, are not wide spread impacting populations.

Legal acts shall be passed as to hold accountable any persons providing falsified, skewed, or misrepresented information that retrieve data, metrics, and/or information for scientific discovery or purposes. It shall be made illegal for anyone to misuse or abuse digital or physical infrastructure and people for scientific discovery. Such violations erode public trust, confidence in science, government oversight, and professional industry operations. Criminal and civil penal codes shall be established to deter research misconduct or industry abuse. Persons in higher authority roles, including research and development, found to have violated their responsibilities, moral obligations, and ethical professional duties to protect patients shall be dismissed without retaliation or oppression, unless evidence proves impact caused harms and/or loss of life. Violators shall be released from duty, after a thorough and comprehensive ethical investigative review with presumption of innocence.

Medical

Intake evaluations will require root cause analysis to identify variables causing diagnosis. Diagnoses shall include assessment with and without the use of technology. For example, consumption inventories shall be completed to assess client thought patterns, habit tracking, peer association and group connections tally, environmental product and service offerings, and nutrition. Diagnosis shall not only be based on scans but include 'process of elimination' checks as to ensure natural healing is prioritized, prior to intensive surgical, pharmaceutical, or bodily intrusions. Patients shall be provided diagnosis determination without the use of technology and technology. First assessments shall be taken with natural remediation given during a period of time. In the event remediation is ineffective, use of -scopy and -graphs for evaluation can be utilized. Artificial intelligence medical determinations shall be disclosed and made separate from medical staff diagnosis. This is to deter lazy medical care that is off sourcing attunement and potentially producing false positive results. Clients or patients shall be informed regarding 'human' doctor diagnosis, 'AI' diagnosis, technology use in care, and any other statistical information inferred from technology use.

Clients and patients are to be educated regarding procedures, protocols, and tools used during their time of care prior to receiving treatments, unless in high risk priority circumstances like ER or ICU. Medical ethics courses shall include patient 'right to know' disclosure learning, integrity materials, and social impacts. Disclosure policies will be mandated to align with transparency and be posted on hospital sites for access, digital and physical. Furthermore, any sponsorship, endorsement, incentivization from third party involvement shall be disclosed. This is to ensure no conflict of interest is at play between pharmaceutical, medical, and/or insurance industries contributing to patient conditioning or dependency on care. Furthermore, pain shall be treated as an indicator requiring deep causal analysis rather than symptomatic suppression with over counter medication. Effectively, care shall not be focused on 'pushing pills' rather than doing sit-ups.

Healthcare industries will be omitted from political lobbying to prevent conflict of interest and bias in public service functions; with the exception of medical integrity officers' inputs in policy research. Integrity officers will not work within private establishments but be given full access to all spaces to evaluate field operations in varying landscapes, collaboratively with other integrity professionals from other industries. This is to assess root cause manifestation impacting persons and populations. Such roles will be protected under law from persecution, retaliation, intimidation, coercion, and any other form of threat due to the nature of the work and commitment to guardianship of civilizations. Integrity officers will be awarded and rewarded for their efforts with free housing, free education for themselves and families, generous financial pay, with honorable designation. This is to promote a culture of integrity and responsibility as well as establish uniform accountability in varying spaces across industry domains. Medical integrity and ethics departments shall be incorporated, if not already in operation. Strict means for confidential reporting shall be made available, including but not limited to paper and/or electronic channels.

Medical professionals making reports of healthcare violations, known to conflict with training, insight and/or intelligence shall be protected under law. Any false representation or reporting shall be a violation of law with imprisonment and fine sentences, as to deter fraud and false statement reporting. The overall objective of founded integrity spaces are to ensure industry alignment to higher purpose function not founded on corruption, misguidance, or greed. Healthcare conferences shall consistently incorporate integrity, ethics, and social responsibility round-tables and speaker opportunities to facilitate discussion and reflection. Moreover, such talks shall have onsite integrity officers to assess and retain expressed feedback for industry improvement. Professional continued learning accreditation shall include opportunities to participate in k-12 school onsite interactive presentations to integrate preventive care content and positive habit routine modeling. This will ensure students lacking home guidance for care taking, like hygiene up-keeping or feminine cycles, are able to be exposed to preventive care material.

Pharmaceutical

Medications, detection kits, devices, and other offerings by the pharmaceuticals industry will need to be evaluated to distinguish if they are benefiting or enabling, reinforcing, or conditioning patients or clients towards risk and/or harms. Furthermore, pharmaceutical devices will need to be taxed as to establish physical and digital structures that prevent illness, ailments, and/or diseases. For example, tax retained shall expand gymnasium development, temples, gardens, and other spaces that educate, character improve, nourish, uplift, and contribute to psychological, social, emotional, financial, spiritual, and physical health. Pharmaceutical product development shall require extensive testing for effects, short and long term impacts prior to deployment or product public launch. Furthermore, pharmaceutical tools and devices shall never be used out of purpose as to hurt or harm populations or carry out national security missions.

Pharmaceutical industries shall be omitted from political lobbying, as to prevent conflict of interest and bias in public service decision making. Except integrity officers' inputs provided forth through policy research as to support evidence based legislation. Pharmaceutical integrity officers will not be able to work within the private sector and be designated to evaluate field operations in varying landscapes, collaboratively with other integrity professionals from varying industries, as to assess root cause manifestation. Moreover, such roles will be protected under law from persecution, retaliation, intimidation, coercion, and/or any other threat due to the nature of their work and commitment to civilizations. They will be awarded and rewarded for their efforts with free housing, free education for them and families, generous financial pay, with honorable designation. This will align to promote a culture of integrity and responsibility.

Supplement, Nutrition, and Plant Based Recreations

Supplements or healthcare promotions shall be founded on the results of tested biology and chemistry metabolic trials presenting short and long term impacts congruent with each species type. No product, in any form, shall be infused with ingredients known to be toxic for public consumption. Any substance or product containing psychotropic altering ingredients shall require permits and licenses, with professionals accreditation and oversight. Accreditation shall include learning in integrity, social responsibility, conscious production, and societal impact. No controlled substance in any form shall be misconstrued, mis-categorized, misrepresented or mislabeled in design via marketing as to entice sale. Furthermore, products containing psychotropic substance ingredients shall not be disguised under 'Plant Based' or 'Medicinal' with false labels aimed to deceive in selling the effects while omitting short and long term impacts on conscience, physiology, and awareness.

Product designs shall make clear, visible, and bold cautionary label on packaging indicating product toxic categories like 'Harmful to Ingest' and 'This Causes Cancer'. Product designers shall have as much joy collaborating and cooperating with compliance and risk departments to ensure nothing is left out, including psychological and social impacts. Any person, from any age, group, or background shall be able to assess a product label and immediately distinguish if it is or not toxic. Furthermore, the supply chain manufacturing channels shall evaluate labels for accuracy and include tertiary testing, as to fault-proof and deter fraud, distortions, biological or chemical infiltration and negligence. Under no conditions or circumstances, any controlled substance is to be made available to the civilian populations or the public that is scientifically known to be neurological and physiological harmful; causing system dysregulation and/or disorder. This is congruent in honoring rights and responsibilities to protect and safeguard, as well as mitigate national security incidents.

Law and Politics

Acts and legislation impacting the ecosystem require time lapse evaluation. For example, laws passed today will require evaluation every 3 years, then 6 years, 9 years, 12 years, and so on; to assess the short and long term beneficial and/or harmful impacts to civilization. This is to identify what has been working and isn't. Mallegislation shall be distinguished from beneficial ones and tracked to ensure enacted acts that weren't helpful or destructive are not incorporated by future generations. Furthermore, such information shall be made publicly available and reasonable accessible within all regional global occupants, as to improve transparency, awareness, civil and civic curiosity and leadership.

Research provided to legislative representatives shall be reviewed by legal integrity officers within each industry as to deter corruption, conflict of interest, fraud, and other misconduct. Furthermore, legal integrity officers shall collaborate with legal officials to ensure public feedback and testimonies are included in research reports and findings. Through randomized selection, research reports and findings shall be evaluated, inspected, and reviewed by independent oversight, not known within the political industries, for ethics and value alignment. Such personnel shall be given a grace period to review pertinent facts and provide sign-offs. Research reports provided for legislation shall not be evaluated on the basis of self-interest, self-righteousness, prejudice and/or bias. As much as possible, each role enacted in political structures shall serve the greater spirit of all people, regardless of origin, across space time continuum. There shall be established channels for public notice and feedback to support proactive improvements via civil and civic change management. In every effort, each nation and planetary region is to be in support of another; in the event of disasters, calamities, difficulties, or hardship. Life is a shared responsibility, and as such, each shall value the life of another as their own.

National leaders, especially those drafting legislation, are to collaborate with other nations to support and serve the development of people across the space time continuum. State divides were not established to divide people for territory, but rather compartmentalize stewarding of responsibilities as to improve life function and operation, aligned to conscience. Politicians, especially those running for office, will need to be allocated a budget stipend by congress to run for office, with funding allocation in scope of needs like lodging, travel, food, attire and beverage. Furthermore, lobbying or corporate sponsorship support shall be omitted from public service role engagements. Business entities and corporations shall only be allowed to support candidacy aligned to constitutional expression like endorsements. Those running for public office will be mandated not to accept any private sector funding; including tertiary methods or loopholes like business operations or fiscal trading. Any avenues leading to corruption, misconduct, exploitation or crimes shall be reported to professional enforcement bodies, like ethics boards and integrity offices for immediate and swift remediation. Misconduct reporting shall be rewarded with honorable designation aligned to integrity for submitting material facts.

Public service professionals will need to comply with ethics reviews, evaluations, and requests. Any relevant documentation shall be submitted timely to support investigations. In the event findings cannot be validated, statements of truth by professionals or witnesses are to be taken under oath and upheld with the presumption of innocence. Judicial systems shall support and accept 'truth' of expression regarding material facts over justifications made by lawyers, second or tertiary statements or data, hearsay, speculation, or technological inferred probabilistic statistical material. It is imperative reports provided to personnel of authority are comprehensive, without omission of facts or pertinent matters, as to not skew judgments and decisions or influence determinations. Justice roles shall have feedback from integrity officers in matters pertaining to legal proceedings involving both civil and criminal cases. Reliance on digital infrastructure, as provided by Artificial Intelligence or brain computing technology, shall be disclosed in processes as to provide subjects better awareness regarding matters and acts. The goal of correction is remediation and rehabilitation. In the event subjects are unaware of misconduct, pertinent information is to be shared, even without the request of discovery. This will further improve societal function, mitigate speculation, and cultural uprising or movements founded on misinformation or misinterpretation.

Enforcement officers paired with community ambassadors, by all means necessary, shall make an attempt to provide educational improvement resources regarding character development and social responsibility to clients; which will no longer be referred to as 'criminals'. Methodologies of 'building connection before making corrections' shall be established. This is to further ensure positive associative stimuli is mapped to the intention, actions, and mission of peace officers and government professionals as well as heal public and civic service trust. Public civilians streaming crime shows shall provide peace officers opportunities to share their incident response stories to ensure comprehensive representation of material facts founded on objectivity. In order to heal current tension stemming from enforcement operations, persons involved in court proceedings can sign-up to share their stories or host social gatherings for improving social rapport and peace.

TV news story spotlights shall include industry professional input like doctors, as to ensure information shared is not framed in dichotomies of 'this vs that' but is authentically comprehensive. This will improve citizen morale, thought patterns of consumers, and cognitive deduction or induction in reasoning. For example, framing stories as 'good guys vs bad guys' amplifies divides; while diverse inclusive dialogue can provide multi-perspective, multi-faceted, and multidimensional evaluation. Furthermore, this shall reflect root cause analysis including those pertaining to anthropology, humanities, and faith. This is not to mitigate or dismiss subject or victim experience, but rather evaluate comprehensively the interconnected processes and systems outside of subject decision making.

Local, state, national, regional, and global enforcement shall serve people as to educate, improve awareness, and consciousness rather than be domineering, oppressive, violating, or cruel in expression or operation contributing to harm or loss of life. Furthermore, conflict facing roles, like law enforcement, shall be awarded more paid time off to improve morale and stress management. This will further mitigate manifestation of profiling or stereotyping propensities stemming from pattern recognition. The foundation of force and power is not found in using it but rather restraint. It is imperative to lead with empathy and cooperation as to uplift and derive intrinsic inspired alignment towards best conduct. Ultimately, true power aims to lift rather than harm. With time, inspired mindful participation will with nourished character development.

Performing Arts

Performing arts industries can integrate education of virtues, value pillars, and principles. As well as, presenting care taking habits that reflect vital health activities and honorable civil social rapport. For example, interweaving teachings in screenplays, scripts, and improv to showcase positive stress coping habits and mechanisms can contribute to optimal health outcomes rather than drunkenness. Furthermore operations like festival with themes that are derogatory to faith produce subcultures that are reactive, resistant to inclusion, and potentially founded on false beliefs. This can exacerbate divides and inhibit critical thought, harmony, problem solving and resilience. Simply, each must practice respect to receive it and be able to set the tone and rhythm for better future becomings founded on emulating kinder function rather than disrespect, hate, or violence.

Establishment of spaces that are diverse can support positive exposure to traditions and culture. In the event a subculture is identified to be operating in harms, it is best for the community and officers to identify and remediate with business owners to address concerns. It's imperative to integrate responsible business oversight of products and services with spontaneous or randomized checks by integrity officers. Social responsibility and accountability is to be prioritized by occupants to ensure safety of all age groups. Any business owner and staff obtaining permits that impact the public shall complete training pertaining to integrity, social responsibility, and wellness.

Financial Sector

True currency has been endowed to each person as reflected with life force energy and time. But due to the evolution of civilization, means to convert energy for compensation and material goods have had an influence on personhood perception. Life was not created with labor conditions as reflected with current economic systems that demand performance and productivity for survivability under the foundation of company brands. It is imperative for all persons to understand and acknowledge that their worth is not contingent on monetary gain, rather character traits and attributes. Furthermore, due to restrictions on obtaining land due to economic wages, persons have resorted to living and renting in high density city condensed areas, where their access to farming is limited. The function of economic structures, especially in high density areas, has contributed to producing dependency on business operations rather than nature. Thus, producing over reliance on economic systems, subjugation on cycles of survivability, and control. To build life, means to support personhood development is required and current dependency may stifle it.

Loans are processed with contingencies that require applicants to meet certain conditions to be offered financial stipend with agreement of funding returns with interest. Due to lived experience conditions, it would serve all, for banks to cooperate and collaborate with non-profit organizations, and other financial institutions that track personhood financial status to ensure fiscal essentials are allocated to each person, especially for children and caretakers. This is to provide each a quality standard of living, so none are in poverty or financial distress or operating in duress as to be enticed to commit crime. It is likely counter-intuitive to have so many divided financial structures with varying financial methodologies of allocating means to retain resources for food, shelter, clothing, and housing essentials; considering persons current freedoms to property and growing food are oppressed in higher density areas.

Mitigating unnecessary financial stress is imperative for improving livelihoods, preventing and deterring crime. For this reason, governments serving their people, non-profit organizations, and financial institutions need consistent methodology for providing access to resources as to obtain essentials and minimum basic needs for themselves and kindred. While governments develop varying programs and innovation causing economic disruptions, it is important to ensure turbulent changes are mitigated impacting access to basic essentials and needs. Furthermore, this can ensure the basic flow and circulation of energy currency is not stifled across terrains and landscapes.

The social impact of integrated monetary operations within the social sphere of experience has yielded many benefits and harms. First, it has produced thought processes centered on the self as it pertains to net worth and value. Second, thoughts have been mapped to things and material gain. Third, monetary effects on decision-making, seen and unseen, alter willingness to act due to assessments of variables contingent on price outcomes. For example, persons may assess funds spent on higher purpose endeavors to be fruitless while investments made in materials to be tangible. This can be skewed in evaluating or perceiving unseen fortune returns that may spur throughout generations. While benefits for having monetary economies allow for appropriation, allocation, accounting, security, and order, there have also been steep costs to social cohesion, harmony, and synergy in such operations. This is reflective in the exploitation consequences based on mapping persons energy to what is tangible as reflected with this 'Morning Star' guided findings. Thus, fiscal economies are reductionist in nature and reallocate life force energy to material possessions and materials. In the short instance, this seems valuable methodology of conversion but due to the evolution of societies it becomes fruitless to trade life force for dirt; as presented with land ownership that was given to all species by design, by their creator. And as the founders of American society have indicated each person has the right to life, liberty, and property.

Amusement Parks and Events

Fantasy and animated motion pictures, reinforced with amusement park development or conferences, and distribution of products may stun maturity development and contribute to manifestations of obsession, over indulgence, or dysegulation. Play is integral for stress management, life pressure decompression, improved learning, social connection, and health; but it should never be exploited as to distract or reinforce consumer cycles founded on risk taking, fallacy, greed or illusions. For this reason, it's imperative to cultivate spaces founded on improving awareness, coherence, and presence. This is not to discontinue attractions or inhibit tourism but to ensure cognitive functions are aligned to improving maturity, perception, and development. Furthermore, this would mitigate saturated stimuli aligned to gratification and forces impacting deduction or induction reasoning towards malintentions and maladaptive actions. For example, this would prevent business monopolization cycles like those founded on scaling motion pictures that trigger behavior towards prostitution via tourism industries and gambling with casinos and entertainment industries Persons who have a strong foundation on fantasy or animation may base their memory, source information, and knowledge on imaginative material content, not founded on honor, energy awareness, or conscience impact. Furthermore, consumption within certain spaces in high foot trafficked areas can allocate energy and financial investments to subset of individuals who keep propagating the same structure operations founded on self-interest like sports leagues.

Hospitality and Business

Hotels and motels require sanitation sweeps to ensure illegal operations aligned with the drug and sex trade, human trafficking, or malactivity are not permissible or enabled; contributing to harmful outcomes impacting societies and national security. Community ambassadors or community response agents shall make regular stops to high trafficked or tourism sites to evaluate premises. Government and business owners shall collaborate, cooperate, and mediate to support community safety alignment founded on integrity and zero violence policies. In the event owners deny or reject compliance, covert checks shall be conducted as to assess business traffic. It is imperative that non-violent or privacy intrusive protocols are conducted to investigate properties. Property owners shall be informed and included through verbal and written communication protocols during site investigations. Transparency will improve and foster trust and safetguards of people, places, and things. In the event operations are harmful, remediation and legal prosecution shall be enacted with respect to due process with sufficient warnings advisories provided.

Community goals shall be established and mediated with community planners and organizers for improving community settings. Awareness in supporting personhood development and growth is to be prioritized, rather than enabling immature and maladaptive behavior. This will enhance energy flow, reduce reactive spending costs, and increase valuations. Community spaces shall be established to support locals in setting goals aligned to sustainable development objectives. Meetings can support community agendas for addressing concerns, manifested harms, and infrastructure deterioration or decay. Government participation shall facilitate, educate, mediate while cultivating safety and driving goal success. Business owners shall establish channels for community feedback, regarding their operations and their impacts. Remediation action plans shall be established when required. While tension and disagreement may arise, providing transparency and long term vision overviews will sooth uncertainty and fears. Ultimately, recognizing the inter-dependency of operation that impact society across planes, dimensions, and landscapes.

Food and Beverage

Designs and labeling of food and beverages shall never be deceptive as to misguide or mislead consumers regarding ingredients. Furthermore, products made in labs shall all be consistently identified with color coded lining near opening site. National leaders shall prioritize the health and well being of all their people, through out the supply chain planning, design, manufacturing, and distribution processes. There ought to be consistencies with product designs, in which groups of any demographics or age groups are able to distinguish and identify products containing toxins from organic or artificial ones.

Education curriculum shall equip students to assess nutrition labels, products, and services both physically and digitally. Such learning material can be reiterative in schools and employer course offerings. Furthermore, they can be in scope of mental, physical, financial, and social health. Proactive deterrence and prevention can be interwoven through the developmental lifespan of each person. Furthermore, this supports improving the quality of life for generational offsprings that improve livelihood and deter harms.

Food and beverage customer service staff can be supported in learning nutritional intake and utilize learning as to support consumer engagement and mindful consumption. For example, while many franchises prioritize selling alcohol upon greeting, it would be best apt to allow service staff to lead with curiosity, acting as consultants, to assess consumers dietary goals then map recommendations or suggestions to them. This can subconsciously prime consumers to be aware of their consumption and align choices to health outcomes. Moreover, this can alleviate impulsive spending and consumption based on moods or emotions and incline mindful consideration congruent with conscience. This can be tested with pilot program initiatives as to asses feedback for improving menu offerings and brand confidence, while aligning to global health goals. Cultivating a culture of conscious consumption, conscious spending, and conscious production through operational changes can kickstart and spark vitality and vigor.

Ingredients in food and beverage shall be assessed and tested for effects, short and long term consequences. Convenience and speed shall not be prioritized above nutritional value and quality. In high traffic venues or arenas, efforts to provide fresh ingredients and offerings like 'watermelon sticks' or 'corn cups' can deter consumption of acrylamide that contribute to obesity or cancer. Moreover, intoxicants drink offerings in high traffic venues can be replaced tasty healthier options like smoothies, fresh coconuts or melons, and flavorful ice.

Retail

Clothing labels and brands require intentional design as to reflect, integrate, and carry higher energy resonance. For example, derogatory slang and absurd graphics can have a negative subconscious influence on demeanor, mood, and expression. Attire printing machines shall require permits for operation, with education curriculum congruent to social responsibility and morale. While some many not participate in aligning attire printing to higher purpose operation, many will be awarded with higher sales through forthcoming mindful cultural integration growing and aligning to responsibility, natural essence, and simplicity. Territory labels, entity brands, and demographic designs can inhibit higher conscious connection while amplifying prejudice, divides, competition, and discrimination. For this reason, it is beneficial to align attire labels, brands, and graphics to neutral design essence to support the awareness of connection.

Insurance

Payments retained by insurance companies, for policy holders, shall allow a percentage of payments funding to be allocated for maintenance costs, repairs, accidents, and replacements. In the event policy holders have made consistent payments over a period of time, each shall be awarded to upgrades when warranted. Moreover, health insurance shall include access to preventive facilities like gymnasiums, botanical gardens, recreation facilities, and healthy food and beverage retailers. This will incentivize policy holders towards physical movement while reducing sedentary inactivity and improve nutritional intake. Furthermore, local community challenges can be hosted to diversify and promote exposure of local cultural business offerings. Rather than funneling marketing through advertisement channels, new pathways to cultivate and foster community can be founded after methodologies for consistent outreach are triggered. This can reduce ecosystem resource demands while improving communication and community participation.

Publishing and Printing

Printing material incorporating messaging or stimuli relative to humanity shall be scoped to responsibility, honor, and dignity as to ensure self-concepts are mapped to optimal conduct rather than be correlated to distortions, illusions, or maladaptive behaviors. Scientific journal materials would best apt be published for printing, rather than be confined to online journals or databases exclusive to members, inhibiting public review. To improve morale, its imperative to support organic and safe material content distribution, as to positively engage and stimulate persons towards best conduct; than feed into distributing chaos, fantasy, darkness and/or drama.

Advertisement and Media

Multimedia has a subconscious effect on consciousness, thorough subliminals, as it projects information to target audience to derive specific thoughts, behaviors, and outcomes. Media and ads require evaluation and assessment for perception distortions, influence, and mental framing impacting worldview and self-concepts, especially those pertaining to affluence and prestige. Such operations can amplify envy or distort function towards vanity.

Digital streaming services have done very well in establishing viewing limits, with incorporated interruption prompts, to mitigate viewing indulgence. To balance viewing offerings, it is imperative to support programming that is honorable and dignified. Furthermore, provide consumers access to content and material that shows positive coping skills and promote better modeling to stressful situations or life events. Media structures influencing personhood towards risk taking and harm, like those amplifying exploitation or violence, need to be deplaformed.

Social platform providing meet-up services between people shall include educational material regarding harassment, dignity, and safety. Such learning can be integrated during the sign-up process, requiring review prior to account setup. This shall provide proactive deterrence and ensure integration of personhood wellness and safety learning. Furthermore, regular notifications regarding safety and wellness shall be made to each account based on digital application usage. For example, if a user is known to have spent x amount of hours, they shall be prompted to complete educational material learning rather than ads. Social platform users are to be provided privacy tips regarding data collection, account cyber security best practices, and other pertinent guidance as advised by the FBI or Cyber Command. For example, relaying advisories notifications regarding romance scam, holiday scams, or pyramid schemes would improve national security and mitigate incidents.

Each digital application shall be category coded. For example, users shall be able to distinguish and identify private corporate apps, government apps, accessory apps, from health apps. Organizing digital infrastructure can support consumers in identifying high risk apps from trusted platforms. Streaming channels, like radio or podcast hosts, shall assess and evaluate topics of discussion with the 'CRAAP Test'. This ensures shared information with the public is current, relevant, authoritative, accurate, and purposeful. This is to ensure shared information is credible and reliable aligned with higher operation in acts and intents. Furthermore, streaming services can assess content with the 'Royalty Test', as to assess tone, language, etiquette, and eloquence of messaging as to align to higher standards of conducts and improve societal culture. Which can reflect and emulate respect, honor, and care.

While young adolescents grow utilizing media in conjunction with play and learning, young adults may continue to mimic and model reality TV shows, drama TV reflective of gossip, and political debates through arguing; rather than unified cooperation and collaboration as to problem solve. For this reason, it's imperative to ensure broadcast medium messaging is aligned towards best conduct and character embodiment that is supportive, resourceful, and helpful. Motion picture producers, screen writers, actors, and other personnel shall not be forced to act in screen plays that jeopardize their morals or values. Each industry shall provide opportunities for meaningful projects without coercion, force, or threat. Moreover, actors shall have appropriate channels to report industry misconduct, confidentially as to mitigate publicity or reputation impact. Producers, screenplay writers, and directors shall be reward for production projects that align to sustainable development goals and improving society health, conduct, and morale.

Music carries frequency vibrations that are processed through auditory pathways via anatomical sensors within the ear canal which stimulates the auditory cortex in the brain. The receiver of sound frequencies, process auditory signals consciously and subconsciously. Tone, harmony, lyrics, rhythm, form, dynamics, texture, speed, and other elements affecting music composition shall be evaluated by tertiary sources. Higher beats per minute can amp up listeners, while lower beats per minute can sooth or calm. These wave patterns can be represented by Hertz frequencies affecting brain waves. Due to this influence, it is imperative persons are aware of auditory impacts and err on the side of caution to energy that impact field coherence.

While the brain waves reflect varying frequencies during waking, sleeping, or meditative states; it is vital to ensure moderation of sensory intake and provide personhood allocated time for stillness or quietness. This can prevent sensory abuse or disturbance that amplify conflict, discord, or violence. Just as the heart has it's rhythm with electric pulse contributing to pumping blood through the body, sound waves can impact brain activity and cognition function. Cognitive operations can be amped up or amp down with movement and emotions. This can impact processing speeds, responses, and decision making. Music distribution and production has evolved since the phonograph. Varying listeners currently have access to diverse music options that are electronically produced. Music used to required a band, with multiple people, to be able to generate multiple rhythmic tunes. However, as of 2025, software programs have allowed the generation and production of music based on singularity. Auditory category selections can be instantly accessed from the palm of the hand via digital devices.

Thus, improving listener awareness can clue knowledge regarding music energy vibration resonance that impact mood, attention, and though patterns.

Risk Management

Risk management teams will need to provide product design and marketing teams with summary reports regarding past, current, and anticipated risks of products and services. Teams will need to integrate the information into marketing, advertising, and packaging labels. The risks will be expressed plainly and reasonably for all populations of diverse backgrounds to understand. In the event designs are overly creative, coded markings can be indented on surface of packaging. For example, an indented line can cue consumers that certain products contains toxins. This integrates multi-sensory processing that is not only visual but also tactile and easily recognizable.

The speed in which information is shared via ads shall be checked by risk managers to ensure coherency is met with volume, decibel standards, and established psychological serving size intakes. This is to align speech to quality for interpretation and comprehension rather than highlighting brand benefits presenting positive effects while speeding through harmful impacts. Products and service offerings affecting neurological responses, especially those weakening the nervous system, cognition, and judgment will need to be restricted from streaming in public media, unless authorized by healthcare practitioners or national security. In every effort, executives and risk managers are to align company or enterprise objectives to civilization development. This will ensure company processes are founded on high purpose operation rather than selfish gain, illusion, or greed.

Military , Intelligence, National Security

Foreign relations, including military tactics and techniques, shall include positive meditative procedures to win over the enemy rather than destroy them. Such could incorporate positive meditations as to build infrastructure related to supporting population development with essentials and basic needs. For example, instead of sending monetary funding to governments who may be corruptible or corrupted, providing necessary tools for agriculture or infrastructure development that supports populous function shall be prioritized in foreign affairs relations. Such investments shall not be considered 'debt' rather gifts to support developing rapport between national bodies founded on values of cooperation, accountability, and responsibility.

Military research and development shall not only include firearms and missile defense training but also ethics, conflict resolution, and cultural awareness learning. Detection and analytic device determinations shall include field assessments, informed witness feedback, and universal guidance as to avoid high innocent causality, oppression, and/or injustice. Under all circumstances, dealing with the military, preemptive actions shall not be activated without due process under the presumption of innocence through time contingent protocol and processes. This is to ensure psychological phenomena like overgeneralization, revenge, or retaliation are not manifested as reactions to incident occurrences that reinforce ignorant function.

Natural resource depletion metrics and standards shall be established to ensure given surface areas are not overused or populated. Each surface area shall support the development of life resting upon it, to be stewarded responsibly. Moreover, native born occupants in a given land, like the British in Britain, shall have primary claims to the region from foreign settlers. Under all circumstances, fostering and cultivating hospitality is the super goal. In the event, foreign visitor settlers are producing violent conflict or harmful outcomes, foreign settlers are obliged to gracefully exit and vacate the area they are in, upon formal request or guidance. Under such circumstances, meditations to support travel shall be funded by the natives with reasonable time contingencies for movement unless founded on prejudice, bias, oppression, or other forms of forced oppression. After a 10 year period, required for de-escalation, new protocols shall be established to support harmonious reintegration and re-entry of foreign travelers into native territories, on the terms and conditions of the natives. At the conclusion of the 10 year period, it is supported to allow for relocation and harmonious re-integration of diverse populations to improve inclusion and diversity, founded on respect and honor conduct within space and matter.

Surface areas and land is to support the development of kindred persons, by keeping families united for enriching and nourishing development while mitigating neglect, abuse, or trauma. Time space continuum can shift interpretation and amplify misconception regarding how things 'ought' to be, for 'whom', and on what 'terms'. It would be best apt to lead with authenticity in evaluating value regardless of location, aptitude, skills, abilities, quantified or qualified measurments. Thus, operating in values and principles can improve cohesion and synergy to ensure areas are supportive of life development, in health and morale.

Under conditions or circumstances where authentication or identification cannot be provided due to damage, destruction, lack of evidence or proof, authentication shall be determined based on lineage artifacts, reasonable witness testimonies, records, and/or higher guidance manifested through signals or intel. Under no circumstance shall any foreigner impart into another territory as to settle or overtake it, without consent or permission. As this violates privacy, sovereignty, established boundaries and life stability. Moreover, global regional negative energy can surface and energetically be made manifest in harms as space time continuum is a magnetic field. Permission shall never be inferred or insinuated without provided written and verbal agreements. With land, to obtained with notarization by involved participants. Signals can be represented in consciousness through personhood expression, which may manifest under specific conditions requiring truth to surface.

In the event foreigners are not able to return to prior land due to coercion, duress, threat, or fear of violence; appeal to natives shall be made through appropriate processes and protocols. In the event persons are seeking shelter or protection from persecution or harm, natives shall provide foreigners allocated spaces. This will allow meditative protocols and investigations to be launched regarding material facts, such spaces can be regarded as 'Immigration Integration Hubs' that support personhood character development, improve cultural learning and awareness. This would be aligned to treating others as a sacred responsibility, regardless of background and circumstances.

It is distinguished that each person has a right to life. If the domicile origin region is oppressive, persons seeking support shall be treated with dignity and respect and be given allocated space to integrate it into without violence. Due to the sensitivity of subjective processing and rationality, space setting and upbringing shall include education regarding social responsibility, cultural awareness , and demeanor training on etiquette. Furthermore, it is imperative national leaders are aware violations towards vulnerable populations are an integrity test. Honoring the meek will support evolving civilization rather than feeding into justified narratives that stem from overt or covert operations. In the event truth or facts are not made manifest, each leaders is required to consult conscience and higher guidance.

In the event foreigners do not have an identified surface area to settle in, they shall attest and express the state of their conditions to higher forces or god and follow provided guidance to new allocated territory. No one person's life shall be deemed more or less worthy than another, based on personal beliefs interconnected to physical merit, awards, status, and other material measures not founded by higher guidance; which devises or derives division, conflict, or violence. Relocation and entry to new land, of any nation, shall be regarded as a sacred entry and opportunity to improve civilization development through fostering peer cooperation and collaboration, relational understanding, and learning towards identifying best aligned conduct. Foreign settlers welcomed into new land shall in their best informed effort support its growth, act therein in integrity, and identify pathways for improving affairs through positive representation, modeling of upright character, and traditions. In addition, participation in communities shall include safe guarding, security, oversight to protection of all people with civility and peacefulness.

Leadership shall be founded on integrity, responsibility, fairness, and honor. In the event leadership of a given region is found to be corrupt or oppressive, other leaders from other upstanding nations shall mediate and serve as interim guides, to support natives with managing responsibilities and duties. Evidence and artifacts attesting to injustice, oppression, or harms misaligned to true justice shall be made to designated integrity office sites across the globe. Corruption, abuse of power, or misconduct reports or complaints shall never be deemed as acts of crime or rebellion, rather as an alert to signal attention regarding areas requiring maintenance for peace, safety, harmony, and synergy amongst a people. In the event such courageous actions are taken in a territory by foreigners, to bring to attention misconduct founded on truth and facts as to improve quality of life, foreigners shall be awarded designation as natives for having proved their commitment and dedication to treating and maintaining the land sacredly as their own. By all means necessary, violence and harmful cruel discipline shall not be enforced. Any violators not utilizing appropriate channels for reporting, like integrity offices or through constitutional intentional expression, resorting to violence shall be prosecuted. All protocols shall handle personhood life with care and dignity, even in detention spaces, regardless of peronshood origin, affiliation, religion, ability or belief. In the event material facts and proofs are ignored, energetic harms manifest through natural disasters cluing that required actions are still required to be taken.

Foreigners acclimating to new cultures shall be given a grace period to learn, integrate, and develop. This grace period can provide intelligence to attest, assess, and investigate claims regarding relocation motivations from past domicile. Inclusion of persons into new territories shall be founded as an opportunity for weighing practices that effect societies. For example, idol worship has been clearly identified through generations through universal feedback to be misaligned to higher purpose living , thus attracting community harms. Therefore, consistent global standardization shall be established within every vertical to ensure idol worship is not being practiced uniformly across continuum, to prevent negative manifestations impacting societies. Foreigners and settlers can come from varying beliefs systems but have energy allocation to support higher conscious operations. Thus, each space is reflective of peronshood intention and actions. When actions and intentions are misaligned, harms likely to manifests. Thus, to improve function, alignment of intentions and actions must be assess pre, during, and post actions and decision making through reflective and introspective processes, procedures, or protocols.

Military industrial tools shall be piloted and weighed for their short and long term impacts and consequences on populations, prior to deployment. Moreover, such testing shall include diverse demographics, identities, personhood representation and landscapes. With all effort, national leaders shall mitigate the likelihood violations to civilian life occur, considering they're responsible for their stewarded populations. Any operations that entrap, entice, and/or reinforce harm in operations, including that of the military due to abuse of power, shall be prosecuted. For example, any falsification or skewed reporting of information of data, metrics, or evidence for biased interpretation as to establish policies or advance agendas serving the best interest of a subset of individuals, purposes, or entities shall be regarded as criminal. Any action that misleads or frauds the public trust or officials will be held as a violation to life and immediately investigated with professional discipline where applicable, under the presumption of innocence. Persons found to have violated their authoritative roles, shall be dismissed to find career opportunities in other fields, without retaliation or oppression unless there was evidence of harm or loss of life. The overall objective in misconduct circumstances is to develop character, rather than assassinate it.

Culture

In an effort to align culture to higher standards of conduct; morale requires subconscious stimulation towards upright character values and virtues. Messaging analogous to virtues, values, character attributes, principles and ideals can be integrated through industry domains. Business operations that distort, inhibit, or mislead individuals towards harm are subject to investigation, regulation, and restriction. In the event subcultures or societies are operating inconsistent to cultivating personhood vitality and vigor, aid to realign shall be made cooperatively and collaboratively. Unless individuals are in such violation are causing loss of life, in which persons shall be detained and investigated instead of continuing harmful operations.

Care taking includes but is not limited to nutrition nourishment intake, cleanliness, hygiene, maintenance, and balancing responsibilities. Care of perosnhood, peers, and the setting or environment will significantly improve morale, health, and safety. Furthermore, care will reduce outcomes of negligence, carelessness, violence, and/or harassment. As subjects apply care taking strategies through actions and behaviors, they'll experience improvements which manifest outwardly. Care, like abdominal core, is a key attribute of character to be exercised and interwoven in decision making and intentions. It is imperative subjects are aware of intentions, surfaced inclinations, and motives as to evaluate drivers, like emotions or unseen forces like scarcity, impacting thought patterns and decisions.

Integrity allows for the integration of unseen positive forces and energy to be enmeshed in the design and development process of life. Persons who make aligned decisions to integrity will uplift their character and collective spirit. Decisions or choices made lacking integrity derive negative outcomes that can be harmful to persons and setting with short or long term impacts on populations and generations. Integrity must be paired with conscience awareness, in an effort to weight surface signals presented or derived from real-time observation. While there is stimuli and external cues that prime perception and influence decision making to urgency and indulgence, it's imperative to infuse introspection and reflection during daily or weekly cycles, to ensure decision processes are aligned to higher purpose operations and integrity. This will mitigate the susceptibility of bias, prejudice, emotional reactions, and unseen forces to replicate and propagate. Furthermore, integrating integrity within decision making can support growing consciousness. In practice, this can assess actions and intents through the 'Good Conscience Test' where subjects ask, "Can I sell this product in good conscience?", "Can I kill this person in good conscience?", "Can I speak on this person like this, in good conscience?", and so on.

Responsibility is having the ability to respond optimally to presented stimuli during circumstances, situations, and/or conditions. To be able to respond optimally, physiological functions require balance and equilibrium. To be support optimal functionality, sensory intake or digestion require moderation and cleanliness. Furthermore, products inhibiting nervous system functions via blockers, stimulants, depressants and others neural mechanisms disruptions can impact awareness, perception, emotions and judgment. This can lead to subjects acting irresponsibly, recklessly, impulsively, and negligently. Conscious awareness will be improved by eliminating irresponsible consumption, production, operation, and function. In having the ability to respond, subjects will likely be able to overcome hurdles with ease and thrive through the life experience.

Dignity is the method in which each person acts on their choices or thoughts. The inverse of dignity is cruelty. Thus, acting in dignity can reflect patience, tenderness, kindness, eloquence, and so on. Relayed communication processes may have developed unconsciously and founded on cultural or societal norms. To improve experiential connection and contentiousness, individuals can apply dignified methodologies of conduct through actions and intentions. Dignified methodologies and treatment will improve diverse cultural cohesion, morale, security, and overall health. While it is hard to break bad habits, once integration is completed new found opportunities to improve culture and international relations will be founded. As Lee Ioacca once said, "If you set a good example, you need not worry about setting rules." Thus, each person can embody and lead their own life to dignity and honor to witness the benefits yielded from operating therein.

Conscious Production refers to intentional forethought in the selection, development, design, assembly, and distribution of energy. Production can include but is not to contributions of food, beverages, events, attires, music and other materialized outcomes from processed stimuli and energy. Production requires contentiousness as to ensure energy is not engrossed with reactive moods, vices, or emotions that lead to maladaptive behaviors. Cultivating conscious production through reflection and introspection slows speeds and provides for improved states that infuse in contributions. This can include slowing down, being aware of impulses, internal signals, assessing intentions, and societal patterns when developing products or services or making legislative changes. Simply, you have to make sure your heart is in the right place. And ensure your personal motivations are not not aligned to greed, violence, hate, or other unseen influences impacting reasons.

Conscious Consumption encompasses the intake of energy through sense processing. For example, this can pertain to food, music, movies, people, products, services, events, and other offerings. Thoughts and decisions can be influenced by internal or external forces, in-congruent with conscious awareness, as to derive specific behaviors toward desired outcomes. For this reason, improving health and vitality requires education, evaluation, and curiosity. Continuous learning and reverse tracing or tracking can equip consumers with information regarding supply chain operations, product development processes or protocols, and owner or founder intentions of materialize outcomes. Some may consider this biblically as 'fruits'. Being able to assess and trace the underlining energy of presented stimuli can improve energy vibration, judgment coherence, and responsiveness. Conscious consumption can reduce hostility and conflict as it requires deeper analysis of material content to assess credibility and validity of material and inter state like those who entice fear, discord, disregulation or violence.

Education

Learning modalities allow educational material to be presented to students as to improve intellectual development, comprehension, and universal understanding. However, learning requires the process of synthesis in order for domain knowledge to be integrated, comprehended, applied, and practiced for deeper awareness. It is important for varying age groups to map learning material to experience for improving understanding. Many recognize students have varying methods in retaining and synthesizing information. Some of these include group discussions that allow real-time in person communication and feedback, accessing and utilizing learning tools, and teaching or mentoring.

As of 2025, technology allows access of diverse resource materials like journals, studies, reports, and datasets. It is important to recognize that students or civilians rely on social support for comprehension due to learning abilities or capabilities, language barriers, cultural backgrounds, and other unseen factors. If the teaching becomes lazy, over reliant on technology and presenting information only via a pane that stimulates some senses, such modalities could stunt learning and widen learning gaps. This is due to students disengagement with identifying how learned materials map to landscape operations, the world, personhood, or universe. Furthermore, this can produce unseen influences on student participation and engagement contributing to withdrawal, loneliness, and digital fatigue. Not to mention, due to over-reliance on digital mediums, students may develop physical ailments pertaining to optic atrophy or eye conditions. Incorporating varying learning modalities in student education curriculum, like scheduling time for synthesis, as well as independent and critical thinking will improve awareness, understanding, health and curiosity.

Students developing through schools may not have foundational knowledge pertaining to the landscapes since they may be migrants or persons of vulnerable households that have changed residents through communities due to home displacement or instability. For this reason, it is imperative to provide practical examples and support presenting learning material of abstract concepts or theories with practical exercises. This can include but is not limited to allowing students to tour production facility spaces, engage in hands on learning and play with physics and chemistry projects, and be able participate in field trips to evaluate earth science and geography. Learning can be fun for all when founded on understanding rather than achievement; which has shown to exacerbate competitiveness and negatively effect of self-conceptual regarding aptitude. Integrating experiential learning opportunities or immersive l earning shall be afforded to all students, regardless of familial household income or learning pace as to ensure civilization development is fair, justice, and equitable. This will improve societies development and civilization as to prevent bias, prejudice, discrimination, and other harmful influences stemming from disparities of diverse backgrounds or upbringing.

Experiential learning shall be integrated in every grade level and equip students with allocated time to synthesis and trace knowledge to the universe. While students are processing through grades with allocated time for retaining and processing information, allowing and including in-person group discussions will foster and cultivate open dialogue and allow for the deduction or induction of multi-perspective processing. This will support the development of empathy and compassion, as many students will be able to gauge the scope of evaluation and worldview of classmates regarding topics. This can further allow teachers, guardians, and caretakers awareness of required flexibility to student learning paces and support student development curriculum.

While some students may come from single parent households or lack support systems, integrating in-person group activities in the learning process will allow students to learn from their peers or engage with learning material differently to improve understanding and awareness. In the event students are only exposed to one form of medium affecting their senses like hearing, sight, or speech; allowing in-person learning with peers will support multi-sensory processing and engagement to retain and synthesize information differently. Knowledge testing should not be based on performance or achievement, as this develops negative sentiment regarding aptitude and amplify competition, rather geared for knowledge retention and synthesis interwoven with curiosity and comprehension. Students will most often need to be reminded of education long term objectives and purpose as they are growing to support curiosity and continued learning participation.

Teachers and guidance counselors shall to the best of their abilities shall explain the objectives, processes, protocols, motives, and intentions of educaiton curriculums to students, at every grade level, as to cultivate positive morale founded on intellectual nourishment, flourishment, and development. This is to ensure gaps of awareness regarding intentions are not misconstrued or negatively interpreted against will or benefit for corporate greed economic greed. School curriculum shall integrate character development courses as to equip students with the understanding for best conduct that will improve their health and prevent illness or disease. Furthermore, teachers, advocates, and professionals shall be provided a generous budgetary salary, separate from classroom appropriation stipend. This is to ensure no undo stress energy is entered through the teaching and learning process. Moreover, education content shall be investigated by education integrity officers across the plane to ensure equitable opportunity to all students regardless of origin or zip code. And such materials in areas are not congruent with negative morale influences founded on illusions, distortions, or harms. While it takes a village to raise a child, providing education opportunities through varying outlets, ensures 'no one is left behind' to thrive and lead a life worthwhile not founded on ignorance.

Government and Licensing

Government was established to aid people towards their best conduct. Enforcement, regulation, legislation, and other sectors require continuous and reiterative alignment checks. Contingencies and mission assessments shall be completed through the verticals as to ensure project and operational intents, objectives, initiatives, and programs are aligned to honor, service, accountability, and transparency. In the event government sector abuse allocated powers and acts therein as to harm or allow harms; persons occupying such roles shall be relieved from their duties and replaced with personnel that support the development of life of all, across the space time continuum.

Government sectors require oversight, independent auditing, checks and balance to value alignment for higher purpose direction. This is to ensure loopholes are not exploited, overreach is prevented, abuse of power is eliminated, and corruption practices are prosecuted. Government sector personnel shall by all means necessary, to the best of their ability, rely and lean into the foundation resting upon the presumption of innocence; while weighing multidimensional and inter-variable evaluation while completing investigations and research. Protocols, processes, tools, systems and other tertiary mediums shall be weighed as third inferences; requiring material discovery, probing, thorough inspection for admissibility. Honor is often brought to government functions when the acts and intentions of the individuals in the sector are aligned to higher standards of conduct that honor all. In the event the macro operation is seemingly unjust, it is likely attributed or reflective of the acts of micro personhood decisions making processes requiring awareness. For this reason, each is to evaluate and heed caution in blaming anyone else for outcomes. This approach in evaluating the self prior to placing blame improves accountability and ensures decisions are not unjust or oppressive. Thus, aligning to ethical behavior and integrity rather than popularity or ratings.

9

Summary

We recognize holding onto historic districts is vital to preserving history. However, we significantly have determined the magnitude and counter-intuitive influence of preserving what leads to hurt outcomes, thus producing a propagation of challenges that absorb time and energy resources in a negative reactive allocation. Moreover, popularizing and misrepresenting harmful activities, products, and areas through deceptive tactics can subliminally reinforce thoughts impacting cyclic behavior and spread manifestations through space time continuum.

Changes to the business life-cycle, operational standardization, technology and legislation will contribute to upliftment by eliminating harms currently impacting the ecosystem through consumption. Changes will target the reduction of cases attributed from ailments and crimes reflected by hospital and prison development and operations; which contributes to 5% of gas emissions. While city income funding may be affected, spending will be substantially reduce. Thus, acclimating to soul enrichment and flourishment as well as freedom from chemical, biological, psychological, informational, and physiological oppression harms, abuse, and/or bondage. Leading and ultimately aligning to true liberty and true justice for all.

10

Code of Conduct

- **Be Curious**: Don't be afraid to ask questions! Inquire when you feel inclined to judge. Lead with curiosity within communication for dialogue improvements, perspective and perception enhancements, and improved comprehension and understanding regarding events, situations, conditions, experiences, etc.
- **Be Considerate**: Throughout your experience, aim to consider other's views, lived experiences, background, inner world, feelings, and other pertinent information before forming an impression or opinion.
- **Be Kind**: The method and tone in how we treat or conduct ourselves can have an improved impact on outcomes. Especially, in uncertain or ambiguous situations. Conducting oneself kindly towards self and others will improve morale, health, and society.
- **Be Observant**: Pay attention to the details! Try to slow down impulses and reactions by practicing noticing skills. Document or log activities to help gain awareness regarding cycles, routines, and habits. Be a witness to support the development of life and civilization, not as a by-standard but a participant and a good Samaritan.
- **Be True**: Honesty is the best policy! In an effort to process signals and stimuli accurately as to make informed decision, being honest and truthful is required of us. This can include sharing pertinent and material facts regarding what we witness and

known information regarding occurrences, internal and external motivations, and intentions. Honesty is time-based, requiring of us to meet the moment bravely with courage. When not aligned to honesty, maladvaptive behavior can manifest like manipulation, deception, omission, and lying that impact responses, inferences, and interpretations; which later ripple in personal and collective experiences.

- **Be Aware**: Awareness allows you to receive and process stimuli and signals in space accurately and appropriately. This can include but is not limited to self-awareness, situational awareness, social awareness, organizational awareness, emotional awareness, and behavioral awareness. Being attuned to inner and external signals can improve conduct, decision making, responses, dialogue, sense making, and cognitive functions. Be cautious and aware of energy intake that interrupt processes. Such may be infused with ingredients that block vital system functions through inhibitions, stimulants, or depressants.
- **Be Conscientious**: Similar to consideration, conscientiousness offers the opportunity to accurately perceive the experience from varying angles. This can be a Birdseye view, walking in someones shoes, attuning to nonverbal cues, or hearing what is left unspoken. Practicing conscientiousness can improve civilization, societal development, and personhood responses and actions; while reducing prejudice, bias, and partiality. In practice, being conscientious can reflect taking the time to process feelings and occurrences as to ensure emotional or knee jerk reactions are not impacting decisions, reinforcing unwanted behaviors, or escalating harmful expressions. Furthermore, being conscientious can refer to honoring and listening to the 'quiet voice' or 'nudges' that surface inside each of us, giving inclinations towards best conduct. This can at times require disengagement with malbehavior like gossip, assumptions, or derogatory speech.

- **Be Compassionate**: Compassion differs from kindness since it allows for practiced mercy and tolerance through actions and intentions. When viewing others in compassion, we can provide ourselves the opportunity to witness and be aware of the human conditions, variables, and complexities impacting personal experiences during development or within the field. This, at times, can reflect extending 'the olive branch' or 'giving grace' in recognizing that others lived timelines and backgrounds may not be all the same. Some persons may have been abused, grown instability, oppressive conditions, or without access to resources like education, moral support, or positive role models. Practicing compassion can support communication, cultivate psychological safety and rapport, improve community connection, and civilization resilience. Acting with compassion demands of us, in strength and willingness, to forgive as to ensure each person we meet is able to experience love, care and inclusion; which later will be emulated in frequency and reverberation.
- **Be Honorable**: Honor is founded on recognizing the soul experience of each of us, is connected to spirit collective, regardless of personhood origin. Each soul in experience is deserving and worthy of dignity. Being honorable is a quality of character that can be developed and aligns with intentions, actions, expression, and motives. Honor is the cornerstone of respect, fair treatment, and true justice. Furthermore, honor or the lack of it is often showcased in our attire and soul garment that leaks into ours expressions, either physically or verbally.
- **Be Faithful**: Often, faithfulness is construed with religion. However, faithfulness requires of us to persevere in alignment to foundational principles, codes of conduct, and ideals like integrity even when it is unconventional, unpopular, or costly to do so. Faith is like a fortress that protects and guards and at times shelters and feeds. Being faithful or of faith instills a commitment to higher intentions and actions, reflected in the vows,

promises, words, commitments and agreements we make. Moreover, it is a force that moves or maneuvers based on conscience and at times unseen guidance. While faith can be traced to belief, it often transcendens it by requiring reliance on the unseen with confidence and conviction; requiring the surrendering of divisions or vices that could harm us and others.

Note

"LOVE NEVER FAILS TO DO THE IMPOSSIBLE, BECAUSE WE ARE POSSIBLE."

Book Soundtrack: The Miracle - Edgar Hopp

Definitions

- **Advertising:** Alternating forms of material content presenting messaging towards a specific target audience with a set intention and purpose. Often advertising is correlated with media marketing presenting corporate products and services.
- **Conscience:** Mental cognition that surfaces before, during, or after impulse or stimulus has been processed by a person. Conscience is an inner navigation system that can support decision making aligned with higher functions.
- **Consumption**: Refers to the ingestion and processing of resources with intention and/or action to allocate energy to desired or specified purposes. Consumption can be based on individual, corporate or entities, national, and/or global operations. For examples, a person may consume food and music through their senses while entities or corporations consume people and materials. Consumption in moderation and conscientiousness will allow for optimal operation rather than mindless with subconscious drivers.
- **Consumption Inventory:** Is through inventory valuation as to assess a consumption, activities, exposures, thoughts, and other operations affecting personhood. This can include food, medications, people, places, media, and thought patterns that develop habits and routine. Consumption inventories are to be completed to evaluate the social, psychological, spiritual, physical, mental, physiological, and financial health.
- **Corruption:** State of mind that influences thoughts of a person and later impacts behavior. Corruption can be distinguished by evaluating priorities, motivations, needs, and influences that impacts thoughts in decision making contributing to manifested outcomes.

- **Culture:** Mental models of norms established by express or unexpressed actions and/or thoughts that impact behavior. Culture often manifests and is impacted by psychology, belief, identity, environment, and other variables in space.
- **Crime:** Crime is an act and/or intention having violated established law within a society. Laws are founded to protect persons, places, and property from harm.
- **Ecosystem:** Landscapes that include plants, animals, and humans. Ecosystems can also refer to organic operation and functions of interconnected systems in an environment.
- **Eradication:** The removal of harm or disturbance causing variable inhibiting optimal functionality.
- **Firewall:** Network device that is integrated to filter traffic. Firewalls can prevent and allow flow as to guard digital assets from attacks.
- **Grooming**: Tactics used by violators to establish rapport and dependency with victims.
- **Innocence:** Encompasses the nativity, lack of intellect, knowledge, understanding, comprehension, and/or awareness by a person as to make conscious and informed based decisions.
- **Ignorance**: Ignoring surfaced signals and/or occurrences through bypassed rationalization and/or justification. Ignorance can escalate to willful blindness when persons deflect, disregards, ignore, and/or avoid presented guidance, stimuli, or intel.
- **Law:** Western law encompasses legislation founded on democratic inferred and agreed upon statutes as to protect people, things, and property. Eastern law encompasses legislation integrated from spiritual and religious protocols. Laws are established based on actions and intentions.
- **Maladaptive Behavior:** Refers to actions taken in response to presented stimuli. Likely attributed from emotional or subconscious states or reflective of mental mechanisms, like cognitive

dissonance, in connection with beliefs, desires, emotions, and/or pressures.

- **Mallegislation:** Legislation enabling to risk taking contributing to harmful outcomes.
- **Maloperations:** Environments that allow enable risk taking and reinforce harmful habits.
- **Marketing**: Is the method a business promotes and/or sells resources like products, services, and events to consumers. Marketing can encompass advertising, operations, consumer and supplier exchange, development processes and systems.
- **Mindfulness**: State of being that allows for the assessment, evaluation, consideration, and weighing of variables and presented stimuli by a person.
- **Integrity:** A valuable mindset supporting optimal conduct of behavior, in seen or private settings, contributing to conscious based intentions and actions. Integrity oriented conduct serves persons, communities, and environment.
- **Porngraphy:** Printed or visual material containing the explicit description or display of sexual organs or activity, intended to stimulate erotic rather than aesthetic or emotional feelings.
- **Responsibility:** Persons acting within an environment have the ability to process and respond to varying stimuli through their senses. Optimal functionality and vitality, produces optimal abilities towards responsible decision making.
- **Root Cause Analysis:** The reiterative process of assessing occurrences through critical evaluation, critical reasoning, and critical thinking as to identify root causes contributing to manifested outcomes.
- **Opioids:** Psychotropic and physiological compounds and ingredients used to influence biological functions for specified desired outcomes. This includes but is not limited to legal pharmaceuticals and illegal drugs.

- **Search Engine Optimization (SEO):** Feature embedded in the website coding that surfaces specific websites, features, or content in search engines to support internet functionality, operations, or performance.
- **Sex Trade:** An industry encompassing personhood sexual exploitation, known as prostitution or sex word. Industry involves sexual exploitation
- **Sociology:** The study of societies and social relations between people, environment, culture, their constituents.
- **Stockholm Syndrome:** A manifestation attributed from victims bonding and tethering to their abusers. This occurs in unbalanced, abusive, relations between child and adult, rich and poor, coach and athlete.
- **Subconscious:** Conscious alert and awareness allows for sense processing and reasoning. The subconscious encompasses the processing of third ques and stimuli beyond conscious reasoning. For example, marketing subliminal messaging operate on the subconscious as to influence consume behavior.
- **Subliminal Messaging:** A form of stimuli that is processed by sensory inputs beyond conscious awareness. Subliminal messaging can impact thoughts, beliefs, behaviors, and effect the function and reasoning of a persons.
- **Synthetic ID**: A digital signature attributed to artificial generated content.
- **Prostitution:** Sexual offers or acts by persons as to retain funding or rewards. Prostitution is an incentivized sexual physical or digital behavior that often causes arousal or excitement to participants or viewers. Persons can be unconscious or diluted and may have been human trafficking victims.
- **Psychology:** The study of mental or psych processing by subjects producing impact on reasoning and behavior. This can expand to include the psychology of on person operating in silo or with others.

- **Webcrawlers:** Computer program bots, agents, or 'spiders' that scan the web with specified intended function. For example, a web crawler can scan a user visited site and create a cache in a database for improved performance or security.

Sources

- Data of Charts and Graphs, Farah Alqattan, November 2024-2025, https://docs.google.com/spreadsheets/d/1kM-sEeY4JvAOiIySyNGW9cCOsN1Mj7zLuHEyIgp-N5rw/edit?gid=0#gid=0
- Monkey, I. (2025, February 3). Top 15 Countries That Supply the Most Pornography Online: Understanding Global Trends and the Role of Ethical Content Creation. Insider Monkey. Retrieved April 22, 2025, from https://www.insidermonkey.com/blog/top-15-countries-that-supply-the-most-pornography-online-understanding-global-trends-and-the-role-of-ethical-content-creation-1441936/
- The Many Faces of Sex Work. (2004). National Institute of Health. https://pmc.ncbi.nlm.nih.gov/articles/instance/1744977/pdf/v081p00201.pdf

  Théry, G. (2016, February). Prostitution under International Human Rights Law: An Analysis of States' Obligations and the Best Ways to Implement Them. CAP International. Retrieved April 23, 2025, from https://www.cap-international.org/wp-content/uploads/2016/11/ProstitutionUnderIntlHuman-RightsLawEN.pdf
- Parliament, E. (2024, September). Regulation of prostitution in the European Union. europarl.europa.eu. https://www.europarl.europa.eu/RegData/etudes/IDAN/2024/762354/EPRS_IDA(2024)762354_EN.pdf
- Novotney, A. (2023, April 24). 7 in 10 human trafficking victims are women and girls. What are the psychological effects? American Psychological Association. Retrieved April 28, 2025,

from https://www.apa.org/topics/women-girls/trafficking-women-girls

- Topical Bible: Prostitution and Adultery. (n.d.). Bible Hub. Retrieved April 28, 2025, from https://biblehub.com/topical/p/prostitution_and_adultery.htm
- Routine Activities Theory: Definition and Meaning. (2021, February 22). Criminology Web. Retrieved April 29, 2025, from https://criminologyweb.com/routine-activities-theory-definition-of-the-routine-activity-approach-to-crime/
- Munoz, A. (2019, September 12). Grooming: Knowing the Signs. The Center for Child Protection. Retrieved April 29, 2025, from https://centerforchildprotection.org/grooming-knowing-the-signs/
- Glasser, W. (2001). Counseling with Choice Theory: The New Reality Therapy. HarperCollins.
  What is a Webcrawler? (n.d.). Webscrape AI. https://webscrapeai.com/blogs/blogsContent/what-is-a-webcrawler-everything-you-need-to-know
- Economics of Prostitution. (n.d.). NCJRS Virtual Library. https://www.ojp.gov/ncjrs/virtual-library/abstracts/economics-prostitution
- McLeod, S. (2025, March 17). Operant Conditioning In Psychology: B.F. Skinner Theory. Simply Psychology. Retrieved April 30, 2025, from https://www.simplypsychology.org/operant-conditioning.html
- Why Sex Work Should Be Decriminalized. (2019, August 7). Human Rights Watch. Retrieved April 30, 2025, from https://www.hrw.org/news/2019/08/07/why-sex-work-should-be-decriminalized
- Whistleblower tells senators that Meta undermined US security, interests. (2025, April 9). The Hill. Retrieved May 12, 2025, from https://thehill.com/homenews/senate/5241043-meta-executives-undermine-national-security/

- LCA: Evaluating Environmental Impact. (n.d.). Acquis Compliance. Retrieved May 12, 2025, from https://www.acquiscompliance.com/blog/lca-approach-evaluating-environmental-impact/
- What Is Social Proof? How to Harness Its Power for Marketing Success. (2023, May 16). Gartner. Retrieved May 12, 2025, from https://www.gartner.com/en/digital-markets/insights/what-is-social-proof
- Monteverde, A. (Director). (2023). Sound of Freedom [Film]. Angel Productions. https://www.angel.com/movies/sound-of-freedom?msockid=1ea21948e9066397234d0ca2e8586202
- Shulman, D. (2024, March 29). Summer of Love | 1967, Music, & Facts. Britannica. Retrieved May 12, 2025, from https://www.britannica.com/event/Summer-of-Love-1967
- Illic, S. (2023, October 26). 12 Countries Where Prostitution Is Legal 2024: Decriminalization vs. Legalization. Southwest Journal. Retrieved May, 2025, from https://www.southwestjournal.com/world/countries-where-prostitution-is-legal/
- Goldenberg, S., Puri, N., Shannon, K., & Nguyen, P. (2017, December 17). Burden and Correlates of Mental Health Diagnoses Among Sex Workers in an Urban Setting. National Library of Medicine, National Center for Biotechnology Information. https://doi.org/10.1186/s12905-017-0491-y
- Zhang, Y. (2020, October 14). A system Hierarchy for Brain-Inspired Computing. Nature. https://doi.org/10.1038/s41586-020-2782-y
- Kahneman, D. (2011) Thinking Fast and Slow. https://en.wikipedia.org/wiki/Thinking,_Fast_and_Slow

Morning Star

www.ingramcontent.com/pod-product-compliance
Lightning Source LLC
Chambersburg PA
CBHW050631190326
41458CB00008B/2225

* 9 7 9 8 9 8 8 1 7 8 5 7 6 *